D1127547

Bioethics in the Age of New Media

Bioethics in the Age of New Media

Joanna Zylinska

The MIT Press
Cambridge, Massachusetts
London, England

For information about special quantity discounts, please email special_sales@mitpress.mit.edu

This book was set in Sabon by SNP Best-set Typesetter Ltd., Hong Kong.
Printed and bound in the United States of America.

Library of Congress Cataloging-in-Publication Data
Zylinska, Joanna, 1971–
Bioethics in the age of new media / Joanna Zylinska.
p. ; cm.
Includes bibliographical references and index.
ISBN 978-0-262-24056-7 (hardcover : alk. paper) 1. Bioethics. 2. Mass media.
I. Title.
[DNLM: 1. Bioethical Issues. 2. Culture. 3. Mass Media. 4. Public Opinion.
5. Technology. WB 60 Z99b 2009]
QH332.Z95 2009
174′.957–dc22

2008041414

10 9 8 7 6 5 4 3 2 1

Contents

Preface

No longer is human existence defined by its unique temporal and spatial coordinates: one body, one life, in a specific space and time. Instead human life is increasingly defined by the agential, instrumental deployment of resources for bodily renewal, both its temporal and spatial context subject to extension or translocation.

Susan Merrill Squier, *Liminal Lives: Imagining the Human at the Frontiers of Biomedicine*

[T]he typical American has come to think of himself less as a citizen than as a kind of patient, whose life purpose is to develop, sustain, and fine-tune his psychological well-being.

Carl Elliott, *Better Than Well: American Medicine Meets the American Dream*

What I want to show is that the general Greek problem was not the *tekhnē* of the self, it was the *tekhnē* of life, the *tekhnē tou biou*, how to live.

Michel Foucault, *Ethics: Subjectivity and Truth*

Bioethics has come to occupy a significant place on the public agenda. Many social groups, encouraged and provoked by the media, have been engaged in an ongoing debate over issues concerning life and health as well as the medical interventions into both: abortion, doping, compulsory vaccination, cosmetic surgery, genetic screening, and so on. However, it is the transformation of the very notion of life in the digital age that has been evoking particular hopes and anxieties among the public in Western liberal democracies. This book engages with many of the ethical challenges that technology poses to the allegedly sacrosanct idea of the human and with the new understanding of the relationship between humans, animals, and machines that new technologies and new media prompt us to develop.

It is perhaps unsurprising that radical technological experimentation should provoke ethical debate. Some commentators have even gone so far as to say, as did Andreas Broeckmann, the director of the major international new media festival Transmediale.05, that the proliferation of new media such as the Internet and cell phones and the developments that have occurred in the areas of robotics and biotechnology have led to "an increased disorientation," which has in turn caused ethical vexation.[1] And yet, even though there has been a lot of discussion among media experts and cultural theorists about the need to address ethical questions—we can mention here the work of Donna Haraway, the projects of Critical Art Ensemble and many of the interventions made at the above-mentioned Transmediale—it is at the level of a permanently deferred obligation that debates on bioethics in the age of new media frequently remain. Indeed, there exists little in-depth theorization of bioethics by cultural and media critics or new media artists. Since the 1970s, mainstream bioethical enquiry has predominantly been the domain of moral philosophy, as well as, more recently, sociology and the health care professions. What has been lacking in most of these more conventional studies is a sustained engagement with media and technological processes, coupled with a second-level reflection on the appropriateness of the traditional philosophical discourses for addressing issues concerning new technologies and new media. It is only over the last few years that media and cultural theorists have joined the debate. Bringing a different set of concerns and theoretical perspectives to bioethical enquiry, they have not so far been very successful in transforming the discipline of bioethics or even substantially shifting the parameters of the debate as such. I hope my book will go some way toward remedying this state of affairs.

One of the principal aims of *Bioethics in the Age of New Media* is therefore to explore and understand the conflicting sentiments in the public domain, as well as among media practitioners and theorists, scientists, and philosophers, when it comes to articulating moral positions about human and nonhuman life and its technological transformations—from the moral panics over how technology is going to destroy biological life as we know it and replace it with its machinic counterpart to a utopian belief in the liberatory power of technoscience. In order to facilitate this exploration, the book provides an overview of what I term conventional or traditional positions in bioethics as developed in moral philosophy and focused more explicitly on health issues. The other major aim of this book is to suggest the possibility of articulating an alternative framework for thinking about bioethical issues. To do this, I draw on some of the current work on ethics by a number of theorists of culture, technology, and new media[2] and connect it with ideas on ethics developed from within continental philosophy.

The "new media" focus of my project requires perhaps some conceptual justification, especially given the terminological instability of this notion. As Martin Lister et al., authors of the eponymous volume on new media, explain, the term refers to, "on the one hand, a rapidly changing set of formal and technological experiments, and, on the other, a complex set of interactions between *new technological possibilities* and *established media forms*."[3] The absolute "novelty" of the media and technologies I am referencing here does not therefore remain uncontested in my book: "the new" always carries a trace of "the old."[4] The "new media" moniker is nevertheless useful as it highlights significant changes—technological, political, and experiential—that have taken place in our society over the last few decades and that have been associated with digitization and the use of computing in so many different areas of life. The term also points to "media and cultural studies" as an interdisciplinary framework that informs my study and its particular set of concerns over the recent years. Media and cultural studies analyze the relationship between humans and technology as a cultural, historically specific process and pay attention to the specificity of the media forms and practices such as digitization, interactivity, convergence, and virtuality associated with more recent technological experiments that shape our ontology and epistemology—our being in the world and our ways of knowing it. The methods, approaches, and texts taken from media and cultural studies are not designed to displace philosophy in my analysis but rather to supplement the more established philosophical debates and narratives when it comes to dealing with ethical questions in the context of the current media culture. I am also aware of the fact that, as the object of my study—human–machine hybrids, genetically modified or cloned organisms, artificial life—is very much in the process of emerging, it is not enough to simply apply the already established theoretical perspectives in order to study it. Indeed, a number of cultural theorists have argued that the forms and practices associated with "new media" call for new models, theories, and approaches that hybridize the traditional theories of media and cultural studies with established philosophies while at the same time remaining open to new trajectories of thought.[5] The book is thus intended, among other things, to stage a dialogue between philosophy, on the one hand, and media and cultural studies on the other.

It is worth remarking that a number of writers who have recently been involved in rethinking bioethics in the context of new media culture—we can mention here Rosi Braidotti, Adrian Mackenzie, or Eugene Thacker[6]—have done so from the perspective of Gilles Deleuze's materialist philosophy. While my book is by no means positioned as anti-Deleuzian, and while it shares intellectual kinship—in spirit, if not in letter—with a number of Deleuze-inspired concepts such as "becoming," "assemblage," or "difference," it also proposes a counterpoint to the more frequent applications in the current scholarship of the Spinozist–Deleuzian paradigm to questions of new media, technology, and the body. In doing so, it brings a very particular framework of "philosophy of alterity" into bioethics, with its primary theoretical references coming from Emmanuel Levinas, as well as two other thinkers whose work on difference inscribes itself in the same tradition: Jacques Derrida and Bernard Stiegler. I want to show with this book that the conceptualization of ethics in terms of hospitality and openness to the alterity (difference) of the other offers a productive model for thinking about life and the human, both in its social and biological setup.[7] As a philosopher who cares profoundly about life, especially about the precarious life of the other, Levinas can provide us with a useful set of concepts for responding responsibly to other bodies and lives as they present themselves to an embodied self. Of course, as many readers are probably already suspecting at this point, the humanist limitations of Levinas's own ethical position will need to be addressed throughout the course of this volume. Stiegler's conceptualization of the human as always already technological, and therefore as responding to an expanded set of obligations, will be decisive in this attempt to make Levinas's ethical theory applicable to what we can tentatively call posthuman subjectivities. Last but not least, feminist interventions into philosophy and new media research play a significant role in my argument, in terms of enabling a reflection on the materiality of the body and reworking the traditional methodology and writing style of philosophy.

As well as being an overview of the existent theories of bioethics, *Bioethics in the Age of New Media* is intended to make a case for an ethics of life rooted in the philosophy of alterity, with life studied in both its molecular and social aspects. This perspective allows me to outline what I term a "nonsystemic bioethics of relations," which facilitates the exploration of kinship between humans and nonhumans (such as animals and machines). It is precisely by pointing to a place of difference that I hope to challenge the hierarchical system of descent through which relations between species and life forms have traditionally been thought. Focusing on the multiple instances in which this difference manifests itself, always differently, helps me to avoid collapsing various beings and life forms into a seamless flow of life. The orientation of my project is thus as much affirmative as it is critical: it has as its principal objective the possibility of sketching out a nonhumanist bioethical framework that does not negate the singularity of "the human" but that interrogates the human's privileged position in the dynamic system of relations with other living creatures and nonanimate beings.

That said, what counts as "living beings" cannot be so easily ascertained. It is through the interrogation of couplings between the living and the nonliving, the human and the nonhuman, that this book aims to challenge the viability of traditional principle-based bioethics—humanist, normative, and universally applicable—in the new media age. It also raises questions not only about positioning the human as a starting point for ethical deliberation around which everything revolves but also about declaring "human life" as an a priori value in all circumstances and as always *worth more* than, say, "animal life." The critique of the position of the animal as a fault line against which the human defines his or her humanity and special place on Earth among sentient beings is part of this questioning. However, rather than shift the analytical focus from humans to animals or even to machines, my analytical unit consists, as suggested above, of *the differential relation between the human and the nonhuman*, with the human emerging via, and in relation to, technology. I do not position technology as something added, then, or as something that disrupts or threatens the original wholeness of the human. Instead, taking inspiration from Stiegler, I understand it as a constitutive and creative network of forces through which the human distinguishes himself or herself from his or her environment. The Greek etymology of the term "technology," in which one can hear echoes of both art and craft, brings to the fore the productivity of the technical relation, which sets up, or creates, the human in the world by differentiating the human from his or her constitutive surroundings—tools, language, memory, environment.[8] This relation between the human and technology is posited here as originary, although it acquires specific cultural inscriptions in different historical periods. New media foreground this relation in unprecedented ways, while also increasing its intensity and multiplying the threads of relationality. Thus, while the book deals as much with the question of technology as it does with the media (new or old), it is through what has become known as *new* or *digital* media that this relationship can, or even should, be evaluated in a new light.

One area where this conceptual transformation has been most evident is the realm of the body. As argued by Bernadette Wegenstein, the body undergoes a process of flattening in the new media age and becomes a screen, a surface of reflection, or, indeed, a medium in itself.[9] In other words, the body is both an image, that is, a medium of communication in the same way pictures, photographs, and magazines are, and "a perceptive apparatus through which the world is processed."[10] In short, it is another medium of communication, alongside television, radio, and the Internet, via which information travels. It is precisely on and through the body as medium—reduced to two-dimensional genetic code, made over in extreme plastic surgery, experimented on in bioart—that different bioethical "problems," or "instances," are enacted throughout the course of this volume.

Departing from the more accepted definition of bioethics as the interrogation of "ethical issues arising from the biological and medical sciences,"[11] I should also make it clear right from the start that my understanding of this concept here is much broader. Bioethics for me stands for an "ethics of life," whereby life names both the physical, material existence of singular organisms (what the Greeks called *zoē*) and their political organization into populations (*bios*). Conventional bioethics has typically been more preoccupied with the *zoē* aspect of life, that is, the "raw," biological life of singular organisms (hospital patients, lab animals, genetically modified species), with an explicit exclusion of *bios*. However, every time we are faced with singular decisions concerning individual beings, and their lives and health issues, we are already situated in, and drawing on, a broader political context. A bioethical decision is therefore not only just moral but also political in its not always acknowledged motivations—but it also has political consequences. (A close reading of selected texts by Michel Foucault and Giorgio Agamben will allow me to analyze the intermeshing of politics with "life" today under the aegis of what the two thinkers term "biopolitics.")

The expanded definition of bioethics as "an ethics of life" I am proposing here raises perhaps a danger of this ethics becoming too overreaching, to an extent that it may be impossible to distinguish between "bioethics" and what we more generally refer to as "ethics." However, I believe that in an era when our bodies and minds are mediated in unprecedented ways, when the Western political concept of citizenship has been medicalized, and when "well-being" functions as a code word for individual happiness and political order, the question of whether there is room for a discrete discipline and approach of "bioethics" needs to be raised. It is the openness and permeability of the traditional idea of bioethics that this book also investigates to some extent. Of course, acknowledging this permeability does not have to amount to collapsing "ethics" and "bioethics." On the contrary, the book makes a case for bioethics as a distinct concept with which ethical issues concerning human and nonhuman bodies and lives, and their interlocking with technology and the media, can be understood in the current conjuncture that is frequently described as the digital or new media age.

In its broad focus on current media and political debates on life, health, and the role and status of the human, *Bioethics in the Age of New Media* is not meant to be just for philosophers. Rather, its intended audience are all those who are concerned about "living a good life" but who are also fascinated, intrigued, or troubled by the instability of what counts as "life," including "human life," in the age of new media, as well as by our constantly evolving relationships with other humans, animals, and technology itself. As the book refrains from postulating what living such a good life ultimately means, it may pose a challenge to those who believe they know in advance what the value of the human and of human life is and who are unwilling to interrogate the meaning and constitution of this value. But it should appeal to those who are prepared to question the humanist foundations of our being in, and knowledge about, the world or who may have already been involved in such questioning via different theoretical standpoints in the humanities and social sciences. The ethical positions with which I align myself most closely throughout the book take issue with deontological moral theories based on a specific content (say, "God," "nature," or "human dignity"). What interests me more are various nonsystemic forms of ethics which dispense with a need for a content-based obligation telling us in advance what we should and should not do but which nevertheless retain a sense of ethical responsibility.

The book has been divided into two parts, respectively titled "Theorizing Bioethics" and "Bioethics in Action." Even though the structure may hint at a theory–practice divide, the overall argument complicates this neat division while also creating problems for the notion of an "applied ethics" which is worked out in advance and employed in particular "cases." Still, part I does provide a more detailed engagement with the classical theoretical perspectives on bioethics, looking at the history of the philosophical debates around the terms "bioethics" (chapters 1 and 2) and offering a possible alternative path for developing bioethical thinking. A number of "case studies"—related to abortion (chapter 1), cybernetics (chapter 2), and the practice of blogging (chapter 3)—already appear in this part, but it is part II of the book, "Bioethics in Action," that presents a more sustained study of different cases and events through which the concepts of "the human," "animal," and "life" are being currently redefined. We encounter there the reconfiguration of bodily identity and national belonging via TV makeover shows and the biopolitical implications of radical plastic surgery (chapter 4), the positioning of the discovery of DNA structure and of the mapping of the genetic code in terms of "cracking the secret of life" (chapter 5), and the utilization of biological material such as tissue, blood, and genes as a medium by a new generation of so-called "bioartists" (chapter 6). However, part II does more than just investigate these three case studies: it also performs the bioethical proposal more explicitly outlined in part I. The bioethics that emerges there takes the form of a framework, or a set of nonnormative pointers which can only be performatively enacted in specific instances. The book thus constitutes an attempt to enact such instances of bioethical intervention under this new framework, through a number of singular events and case studies. In all of these case studies, the notion of "life" is being reconceptualized in a radical way. Bioethics thus presents itself as a problem in the making, requiring first of all a theoretical and critical vigilance rather than a definitive solution. It is precisely a departure from the model of bioethics as "a technical fix" to a moral problem that distinguishes the bioethical project undertaken here from many of its more conventional predecessors.

The book ends with a call for "being-in-difference," which is to be seen as a hospitable—if not uncritical and unconditional—opening toward technology by a rational sentient being, but also as a form of bodily passivity, of letting oneself be-together-with-difference, or of being-mediated.

Acknowledgments

Many people have helped me develop and fine-tune the ideas contained in this book. I especially want to thank Clare Birchall, Dave Boothroyd, Rosi Braidotti, Dorota Glowacka, Gary Hall, Sarah Kember, Angela McRobbie, Mark Poster, Nina Sellars, Doug Sery, Stelarc, Bernadette Wegenstein, and Adam Zaretsky, as well as the anonymous MIT Press reviewers. I presented parts of the book as talks at the University of Amsterdam, Australian National University, Brown University, University of California, Irvine, Cambridge University, University of Edinburgh, Free University Berlin, Lancaster University, Johannes Gutenberg University in Mainz, University of King's College in Halifax, University of Melbourne, University of Reading, Utrecht University, Thames Valley University, and University of Western Sydney. I received generous hospitality and a lot of helpful suggestions from all these places. An additional thank-you is owed to my students on the MA Digital Media at Goldsmiths, who served as guinea pigs for testing many of the ideas that made their way into this book.

I

Theorizing Bioethics

1

Bioethics: A Critical Introduction

We Are All Bioethics Experts Now

By way of opening this chapter, I would like the reader to consider a few (rather big) questions:

- Is abortion murder?
- Should one be able to buy a kidney if one's life is under threat and one can find a willing seller?
- Do we need to be concerned about the possibility of cloning humans?
- Is experimentation with animals morally permissible?
- Should doctors always tell the truth?
- One what grounds, if any, would one object to organ transplantation from pigs to humans?
- Is gene therapy an attempt to produce a new "master race"?
- Is any life worth living?
- What is the value of human life?

One does not need to be a trained philosopher to attempt to answer these questions. Indeed, most people would probably be able and willing to provide an answer to at least some of the above—even if these answers were to amount to mere opinion or something like "I'm not really sure" or, perhaps, "It's rather complicated." When it comes to matters concerning our life and health, there seems to exist an unwritten consensus that they must not be left just to experts—philosophers, theologians, or doctors—and that all freethinking citizens in liberal democracies need to have a say when decisions are being made about their lives and bodies. Of course, not all such answers will be philosophically astute; some will consist in a mere repetition of the most orthodox views developed by religious or secular experts. However, it is the very possibility of participating in the discourse on human life—a discourse whose signal points are being increasingly tested by technological developments and experiments—that is important here and that is being claimed as a right. What one will specifically say in response to these questions depends on one's intellectual and moral position: on what concept of human life one subscribes to, whether it is underpinned by religious or secular viewpoints, whether life here is seen as a superior value. One's responses will also depend on one's political convictions and one's understanding of issues of property, freedom, and social justice. This is not to say that these responses will be fixed forever. The very process of decision making is potentially dynamic, in the sense that one's values and convictions may undergo a transformation when exposed to new moral problems and questions. As new technologies and new media are constantly challenging our established ideas of what it means to be human and live a human life, they also seem to be commanding a transformation of the recognized moral frameworks—although this is not to say that the need for such a radical reassessment of values is taken as a given by everyone.

And yet debates on human life, health, and the body are never just a matter of individual responses and decisions made by singular moral entities. Instead, they belong to a wider network of politico-ethical discourses that shape the social and hold it together. The broadcast media, with their moral panics about "Dr. Dolly" attempting to clone humans or about asylum seekers infecting the home population with serious diseases such as TB and AIDS, as well as their more considered reports into the mismanagement of mental health care in the United Kingdom, play an important role in constructing narratives on human life, health, and the body.[1] No matter what position is actually taken, it is the very possibility of participating in the discourse on human life that is important here and that both media producers and media audiences are claiming as a right. Thus, even if, as stated earlier, most people can be said to "have an opinion on life," I am principally interested in how certain positions and opinions on life become legitimized as authoritative and hegemonic. In other words, I want to explore the emergence of the academic and professional discourse known as "bioethics" that has framed and legislated the debates on life and its technological mediations and transformations. Arising in conjunction with, and in response to, developments in the areas of biotechnology and medicine, bioethics raises philosophical questions about the constitution of the boundaries of the human and human life, as well as considering policy implications of such developments for government bodies, health care institutions, and other social organs. It is thus always already a clinically driven "expert discourse," which can then be applied to "real-life cases." However, bioethics is also an academic discipline, underwritten by the disciplinary procedures of moral philosophy (although theology and sociology also contribute to its intellectual trajectory). Originally positioned at the crossroads of the clinic and the philosophy department, bioethics has in recent years attracted the attention and investment of "Big Pharma," that is, the biotechnological industry.[2] In spite of the differences between the European and American bioethical traditions, we can risk saying that globalization and the financial investments into medical and ethical research programs by international biotech companies have strengthened the Americanization of bioethics across the globe over the last decade or so.[3] Globalization and neoliberalism have also pushed the utilitarian agenda of this newly emergent "international bioethics" much more to the fore.

Philosophically, mainstream bioethics most often employs deontological perspectives and attempts to prescribe universalizable judgment for all possible circumstances, as explained by Helga Kuhse and Peter Singer in the Introduction to their anthology, *Bioethics*.[4] It is thus a form of applied ethics, whereby general rules are applied to different cases. Bioethics frequently adopts the philosophical framework of utilitarianism, involving the methodical calculation of goods under given sociopolitical circumstances in order to satisfy the greatest number of desires and preferences. Ethics here is not a matter of taste or opinion; instead, it is amenable to argument—and indeed, from this perspective it is the responsibility of thinking human beings to engage in argument. For this argument to be productive, consistency and factual accuracy need to be ensured. Other philosophical positions that Kuhse and Singer list as playing a significant role in bioethics today involve a Kant-inflected belief in the inviolable moral principles formulated in the categorical imperative; the Aristotelian ethics of good based on certain adopted views of "human nature"; Christian ethics of natural good and evil regulated by the idea of God; and, last but not least, ethical positions that are not based on any principles or rules but rather on an idea of what it means to be a "good person" (and more narrowly, a "good doctor," "good researcher," or "good academic"). What binds all these different positions on bioethics together is the following:

- the sense of normativity they all embrace, which is filled with positive content, that is, the idea of good they refer to and defend;
- the rational human subject that can make a decision and that is seen as the source of this decision;
- the need for the universalization and applicability of the moral judgment.

It is on these three counts—predefined normativity, human subjectivity, and universal applicability—that I want to raise questions in this book for what I broadly refer to as "traditional bioethics." The aim of this chapter is therefore to present an overview of dominant positions in bioethics as developed from within both moral philosophy and health-related professions, while also considering the financial and affective investments that underpin those positions. This overview will prepare the ground for our consideration of the possibility of thinking differently about the life and health of individual citizens as well as whole populations in what I have tentatively called "the age of new media." I will outline—here and in other chapters—a number of such alternatives which have recently been proposed by thinkers who have remained attentive to technological processes at all levels of life, such as Rosi Braidotti, Rosalyn Diprose, Carl Elliott, Donna Haraway, Margrit Shildrick, or Eugene Thacker (to name but a few). I will also offer my own contribution to these debates.

The majority of these alternatives in thinking about bioethics inscribe themselves in a broader set of debates between foundational and nonfoundational, systemic and nonsystemic, or—to resort to something of a cliché—analytical and continental traditions within philosophy. Inspired by the hybrid that in Anglo-American academe has gained the name of "continental philosophy" (I am referring here to the predominantly French and German-influenced approach that posits reality as always already in need of interpretation and historical contextualization, rather than a timeless logical structure in need of clarification),[5] they are also informed by interdisciplinary work on ethics within media and cultural studies, English and comparative literature, and sociology.[6] My own line of thinking, as well as that of many of the other theorists I will be drawing on here, arises as a response to deontological moral theories which are based on a specific content (i.e., good that transcends Being in Plato, the almighty and all-loving God in Christianity). What is put forward instead is a nonsystemic ethics that dispenses with a need for a content-based obligation, while at the same time retaining the sense of duty (i.e., the concept of the obligation to the other in Levinas and the notion of active production and the expansion of life to its full potential in Deleuze).[7] The specificity of my own argument lies in bringing the Levinas-inspired understanding of ethics as responsibility for the infinite alterity (i.e., difference) of the other, as openness and hospitality, to debates on bioethics. However, this understanding of "open-ended" ethical responsibility is also underpinned for me by a cultural studies injunction to study, attentively and singularly, multiple instances where responsibility imposes itself against specific forces and powers acting in the world and where it requires a careful negotiation with contradictory claims for such an openness.

Before I move on to outline any such alternatives, though, I would first like to spend some time examining further some of the main principles of traditional bioethical theories, focusing on their philosophical premises and political underpinnings.

"Traditional" Bioethics and Its Discontents

Kuhse and Singer explain that the term "bioethics" "was coined by Van Rensselaer Potter, who used it to describe his proposal that we need an ethic that can incorporate our obligations, not just to other humans, but to the biosphere as a whole."[8] Although ecological concerns are not foreign to many bioethicists, nowadays the term is used "in the narrower sense of the study of ethical issues arising from the biological and medical sciences."[9] A branch of applied ethics, bioethics is most commonly seen as requiring the formal logic, consistency, and factual accuracy that set a limit to the subjectivity of ethical judgments. In most cases, however, the requirements of formal reasoning have to be reconciled, in one way or another, with "practical constraints." Kuhse and Singer postulate "universal prescriptivism"—prescribing universalizable judgment for all possible circumstances, including hypothetical ones—as a promising alternative to both ethical subjectivism and "cultural relativism." They explain, "The effect of saying that an ethical judgment must be universalizable for hypothetical as well as actual circumstances is that whenever I make an ethical judgment, I can be challenged to put myself in the position of the parties affected, and see if I would still be able to accept that judgement."[10] Judgment is thus being made by a rational, self-enclosed and disembodied self which remains transparent to itself and which can extricate itself from its custom and culture, that is, its *ethos*—a point to which I will return later on in this chapter.

This ethical position has been developed by the Oxford philosopher R. M. Hare and is known as "consequentialism," a form of utilitarianism which is based on the view that the rightness of an action depends on its consequences. We can hear in this position echoes of Kant's moral philosophy. For Kant, morality has to come from our reason, rather than from any external concept of good, and it does not involve any principles that would not be subject to universalization. His categorical, universal imperative finds its application in the so-called "Formula of the End in Itself," which demands that we treat "humanity in your own person or in the person of any other never simply as a means but always at the same time as an end."[11] Postulating respect for other persons, Kant's ethics stems from the (rational) self which is *naturally* conducive to moral judgment. While a number of contemporary consequentalists, including Hare, are more interested in "practical" resolutions to moral dilemmas, for Kant there are inviolable rules which cannot be changed even if the moral majority would like them to be adjusted in one way or another.

Universal prescriptivism as promoted by Hare, Singer, et al. is not based on any notion of a pregiven universal good, but rather on what we might term the methodical calculation of goods under given sociopolitical circumstances. In this way, Jeremy Bentham's and John Stuart Mills's utilitarianism, whose ethical principles were aimed at ensuring the "greatest surplus of happiness," is modified: the idea of maximizing the net sum of all happiness is abandoned for the sake of a more modest attempt to satisfy the greatest number of desires and preferences. (Neo)utilitarian positions of this kind inform a great number of debates among contemporary bioethicists. Kuhse and Singer's own ethical proposal, rooted in utilitarian philosophy, goes beyond any predefined rules, no matter if drawn from reason, human nature, or God. It also puts in question the teleological explanation for ethical laws. If humans are seen as purposeless beings who are the result of natural selection operating on random mutation over millions of years, "there is no reason to believe that living according to nature will produce a harmonious society, let alone the best possible state of affairs for human beings."[12] Instead of a priori rules, Kuhse and Singer propose practical solutions. However, when they explain admiringly that utilitarianism "puts forward a simple principle that it claims can provide the right answer to all ethical dilemmas" and that can be applied universally, they position ethical quandaries as disembodied and decontextualized technical problems that concern singular subjects in isolated circumstances.[13] Bioethics becomes here a "technological fix" to a technical problem.

A similar view is espoused by Stephen Holland in his Introduction to *Bioethics: A Philosophical Introduction*, a book that presents an account of positions in bioethics which are rooted in analytical moral philosophy. Holland states there that "a grasp of normative moral theory is required to address practical ethical problems."[14] This statement clearly foregrounds the view of ethics as expertise rooted in predecided moral norms that can be applied to specific cases. And yet it can be argued that this kind of approach to bioethics and, more broadly, "life itself" risks turning ethics into an automated program that is somewhat schematically applied to specific cases, without taking too much account of the fact that the cases themselves are still very much "in the making." Indeed, in encounters with new technologies and new media, the ideas and material forms of the human, the body, and life itself are undergoing a radical transformation, with new forms of kinship between humans, animals, and machines being constituted and with the human itself being repositioned as "a digital archive, retrievable through computer networks and readable at workstations."[15] This is by no means to suggest that the human has been reduced to information in the age of new media and that we can therefore do away with embodiment; it is only to point to the emergence of new discourses of the human which undermine its centering around some fixed biological characteristics or moral values. "Applied bioethics," understood as the application of the previously agreed moral principles, informed by rational argument and based on biological knowledge, can thus perhaps be seen as threatening to close off an ethical enquiry into the emergence of, and encounters between, organisms and life forms that defy traditional classification all too quickly.

Another problem concerning bioethics which is rooted in the formal reasoning of moral philosophy is that it often relies on hypothetical case studies which function as intellectual exercises but bear little relation to the actual, material circumstances resulting from the developments in biotechnology and new media (no matter whether a case is being made in support of, or against, issues such as abortion, xenotransplantation, or gene therapy). A frequently evoked example is the one put forward by Judith Jarvis Thomson, whose proposition that abortion is morally defensible is derived from the invocation of the figure of "a famous unconscious violinist" who has a kidney disease and has been connected to another human being for nine months in order for his disease to clear.[16] The case is supposed to exemplify the excessiveness or even ridiculousness of a demand posed by a supposedly worthwhile human being—a violinist but also, by extension, a fetus—who, by nature of his or her special talent, and the future potential to which it can be put, has the right to take away the freedom of another human, without considering the latter's consent or well-being. We are faced here with a philosophical argument constructed through analogy, whereby the specificities of different situations and cases are eliminated. Again, calculation becomes a dominant tool in this kind of moral reasoning, with different a priori principles being weighted against each other in an attempt to decide whether they are broad or narrow enough.[17]

Human, All Too Human

All the issues listed above notwithstanding, it is the inherent humanism of much of traditional bioethics, be it in its religious or secular form, that I find most problematic in contemporary bioethical thought. Let me illustrate what I mean by this by continuing with the abortion example. In his contribution to a debate on abortion, John Finnis, an expert in jurisprudence and constitutional law at Oxford University and author of many books on natural law, fundamental of ethics, and moral absolutes, writes:

> Leaving aside real or supposed divine, angelic and extraterrestrial beings, the one thing common to all who, in common thought and speech, are regarded as *persons* is that they are *living human individuals*. This being so, anyone who claims that some set of living, whole, bodily human individuals are not persons, and ought not to be regarded and treated as persons, must demonstrate that the ordinary notion of a person is misguided and should be replaced by a different notion. Otherwise the claim will be mere arbitrary discrimination. But no such demonstration has ever been provided, and none is in prospect.[18]

Finnis's condemnation of abortion is based on the principle of "active potential" embraced by many bioethicists—a belief that the embryo "is a human being and human person with potential, not only a merely potential human person or potential human being."[19] The embryo is thus perceived as a "human individual from the beginning of fertilization."[20] The ontological status, universal meaning, and transcultural value of "the human" (or, indeed, a "living human individual") is presupposed in this theory as a given. Significantly, the very same argument based on what we can describe as a "stretched scale of personhood"—from a potential human being through to a human being with yet-unfulfilled potential, and then to a human being whose potential is being realized to its maximum capacity—is used by Finnis's opponents. For example, the philosopher Michael Tooley outlines his defense of abortion by postulating "a basic moral principle specifying a condition an organism must satisfy if it is to have a serious right to life" and then arguing that "this condition is not satisfied by human fetuses and infants" and thus that "they do not have a right to life."[21] The sliding scale of humanity and personhood is being applied in both types of moral argument. It is the positioning of the object of bioethical enquiry on this scale that determines the moral response to it.

Interestingly, a certain opening seems to have been created in Tooley's argument when he calls for a need to distinguish between a human being and a person, with only the latter being moral or having moral rights, including the right to life. "Person" thus becomes for Tooley "just" a moral concept, a tactical maneuver synonymous with asserting that X has a moral right to life,[22] raising the possibility of developing a nonhumanist, rights-based bioethics—if only we could agree in advance what it actually means to be alive. However, this is an impossibly big "if." The distinction between brain death and cardiac death introduced over the last few decades in medicine and the reconceptualization of life as emergence and evolution by researchers in computing and artificial life have cast doubts over the certitude of our all-too-human understanding of the concepts of "life" and "being alive" (even if the alife discourse ultimately reinforces the humanist assumptions it sets out to challenge).[23] Significantly, the author of *The Birth of Bioethics*, Albert R. Jonsen, informs us that the key question bioethics grapples with concerns precisely the ontological status of the human, and human life and death, with, for example, Robert Morrison defining death not as an event but rather as a process commencing at the beginning of life and progressing through its entirety, and Leon Kass postulating that death is an event which should be defined by specific physiological criteria.[24] The possibility of the critique of humanism, and of the inherent "truth" of the human and its preestablished, albeit competing, definitions of what it means to live a meaningful life, thus presents itself as inherent to bioethical enquiry. Coming back to Tooley's proposition, even though the identity of the person presumed by him is strategic, its humanism is nevertheless asserted by a somewhat hesitant aside: "it seems to be a conceptual truth that things that lack consciousness, such as ordinary machines, cannot have rights."[25] We can see from the discussion above that Finnis and Tooley prioritize pragmatic solutions over speculative debates. They do indeed consider a possibility, somewhat jokingly or hesitantly, of the existence of other life forms, "real or supposed divine, angelic and extraterrestrial beings," only to position these beings as exclusions, concepts that should not detract a moral philosopher from the serious task of interrogating an already established person's rights or intrinsic value. And thus the inevitable question, "What about out-of-the-ordinary machines?," that many a theorist of technology and new media would like to pose to Tooley, remains unanswered.

Significantly, even Peter Singer himself—a veritable enfant terrible of contemporary bioethics due to his unabated support for euthanasia and the killing of anencephalic babies (i.e., babies who have no cerebrum or cerebellum but only a brain stem)—resorts to this very same "stretched scale of personhood" when outlining his ethical propositions. In *Rethinking Life and Death: The Collapse of Our Traditional Ethics*, Singer introduces, in a similar vein to Tooley's argument, a distinction between a "human being" and a "person," with only the latter, characterized by rationality and self-awareness, being worthy of ethical respect. Singer includes nonhuman animals such as great apes in the category of "persons" and believes that "whales, dolphins, elephants, monkeys, dogs, pigs and other animals may eventually also be shown to be aware of their own existence over time and capable of reasoning."[26] While his "new ethical outlook" raises radical questions about the principle of the sanctity of human life, his concept of the "person" only extends the notion of the human as a rational being worthy of ethical respect. For Singer, the "new humans" are still skin-bound, carbon-based singular entities, and thus his bioethical propositions are merely an expanded version of traditional moral theories. Although Singer does encourage his readers to interrogate the boundaries of life and death, he does not really investigate the philosophico-political model (i.e., the political philosophy of self-interest and possessive individualism) which underlies his notion of the human. Indeed, not much recognition is given in his work, for example, to the fact that life sciences such as biology and primatology, rather than being just a mirror reflection of capitalist social relations or gender structures, actively reproduce them.[27] In Singer's moral universe there is no room for a thorough investigation of the intermeshing of wider political processes and cultural influences with moral dilemmas.[28] What he therefore ends up proposing is an ethics of (and for) the individual, who has to make rational moral choices as if he or she could always be carved out from the network of relations and flows of capital. Nor does it occur to Singer to include an investigation into the antagonisms that organize the social: any analysis of wider sociopolitical processes seemingly needs to be separated from moral judgments. In his theory of bioethics we are presented with a rational working out of rules, a process of calculation where values can be compared for the sake of elaborating a common good.

While obviously radicalizing humanist ethics by shifting the boundaries of who counts as a "person" (an ape or possibly a dolphin may, while an anencephalic baby does not), Singer still preserves the structural principle of this ethics, with an individual person serving as its cornerstone. In Finnis, Tooley, and Singer, then, all of whom I have included in this chapter as representatives of radically different moral theories, both the moral agent and the object of bioethical enquiry are defined as singular self-enclosed entities, extricated from the networks of social relations and political circumstances as well as the material and discursive conditions of their own emergence. In religious and also secular versions of many bioethical theories, bioethics conjures up the idea of a freethinking neoliberal subject, both as someone who is in charge of making a decision and someone upon whom a decision regarding life and death is to be made. Finnis's fetus is a potential person, which is why humans as rational moral subjects have a responsibility to make this decision on their behalf, in order to enable the realization of their personhood, while Singer's apes and dolphins are perceived to be "like humans" and therefore deserving person-like moral treatment.

Even the British moral philosopher John Harris, an unabashed supporter of "human enhancement" and a stringent critic of social hysteria over any type of alteration to humans' mechanical or chemical make-up, turns a blind eye to the sociocultural circumstances of his technologically enhanced moral subject and thus ends up reaffirming its humanism. In *Enhancing Evolution: The Ethical Case for Making Better People*, Harris posits the need for enhancement as a universal "moral imperative" and seems to have a very clear sense what this "enhancement" actually means. His "better people" will be more intelligent, more beautiful, but also "longer-lived, stronger, happier, smarter, fairer (in the aesthetic and in the ethical sense of that term)"—in other words, "more of everything we want to be."[29] While I am in agreement with Harris that there is no need for a moral panic over enhancement since "many of us are already enhanced," there is absolutely no realization in his argument that the allegedly objective human qualities he presents as desirable are actually cultural values, underpinned by numerous assumptions and judgments. What is more, the issue of equal opportunity, which is the guiding force behind his project, cannot be resolved merely on a philosophical level the way he proposes, without addressing the broader questions of politics and its alleged progressivism, which Harris seems to take for granted (in the sense that the "good" of enhancement enjoyed by the "early adopters" will then spread into whole populations), or the logic of capitalism in which, arguably, a certain sense of inequality is imbedded. To think that technological enhancement as such will magically solve the issue of inequality is not particularly innovative—various technolibertarians have thought that about the automobile or the Internet—but it is politically reductive and hence rather naive. This is precisely why cultural studies, which has a long history of thinking through the interconnections between culture, politics, and "the individual," could teach many a moral philosopher a lesson about the structurations of power and the impossibility of a neat separation of entities for the sake of an elegant moral argument.

Supported by the logic of "stretched personhood" which nevertheless posits the person's boundaries as fixed, the bioethics that develops firm moral positions in advance and then applies them to specific cases may therefore be difficult to retain if the self-enclosure of "the person" which is its prerequisite is revealed to be both a philosophical and a biological fiction. A number of examples which stretch or enhance individual personhood in totally unpredictable ways, perhaps even beyond the point at which calling them "human" is still applicable, could be evoked here. If we take into account the radical opening of the boundaries of the human body and life—through prosthetic enhancements such as corneal implants or gene therapy, programs such as the Human Genome Project, and the redefinition of death through the notion of being "brain dead"—the presumed humanism of what I call here, for reasons of brevity, "traditional bioethics" is found wanting. However, I want to suggest that a more fundamental reconceptualization of "enhancement" is needed.

Both experiential and theoretical developments in the areas of new technologies and new media are calling on us to radically rethink the mainstream understanding of technology as a tool that can be applied to discrete entities. It would be more productive perhaps to envisage instead a mutual coconstitution between the entity that gets designated as "the human" and its technology. In other words, if we think technology beyond its Aristotelian concept of a mere tool and see it instead as an environment, or a field of dynamic forces, we will have a more interesting and more critical framework for understanding "human enhancement" or "extension," with prosthecity being repositioned as an originary relation between living and nonliving entities.[30] This repositioning will also allow us to analyze the political vector of many of these forces as well as their material consequences.

It should be mentioned here that a critique of traditional standpoints in bioethics which focus on a disembodied rational subject removed from its sociopolitical circumstances has been ongoing among a number of feminist philosophers. To return to the earlier abortion example, through which I highlighted similarities and differences between its opponents (Finnis) and supporters (Tooley), the feminist philosopher Laura M. Purdy defends abortion on the ground that the personhood of a woman is erased in most anti-abortion positions focused on protecting the unborn. As a consequence, such positions reduce the woman to a mere "fetal container."[31] Although Purdy does argue for the need to examine the social and economic context in which decisions about women's bodies and their health are made, her standpoint is still rooted in liberal philosophy, whereby the pregnant woman is treated as an individual moral agent with rights. The "average white middle-class man in the street" remains here a measuring stick against which ethical injustice carried out against women is judged. Many other feminist positions in bioethics adopt a similar (hu)manist perspective. For example, the International Network on Feminist Approaches to Bioethics (FAB), founded in 1993, focuses on developing a more inclusive theory of bioethics which encompasses the standpoints and experiences of women and other marginalized social groups. FAB also examines presuppositions embedded in the dominant bioethical discourse that privilege those already empowered and attempts to create new methodologies and strategies responsive to the disparate conditions of women's lives across the globe.[32] While the significance of such feminist approaches to bioethics needs to be appreciated—they have been crucial in challenging the orthodoxy of many law-making bodies, changing the discourses and practices around health care, and ensuring more equality for women, people of different races, the disabled, and homosexuals—FAB's dominant agenda nevertheless conforms to a large extent with the humanism which underpins most of moral and political philosophy. It does this by focusing on women and other excluded groups as moral agents with particular experiences of oppression and particular identity-based standpoints and "voices."

There have been attempts coming from other feminist theorists to radically rethink the liberal human-centered framework underpinning bioethical debates: one can think here about the work of the aforementioned Rosalyn Diprose, Rosi Braidotti, Sarah Franklin, Donna Haraway, or Margrit Shildrick (to name but a few). However, before I move toward sketching a number of such different feminist propositions on offer, I want to interrogate a little further some of bioethics' more conventional aspects and legacies.

A Medical History of Bioethics

While the sections above have dealt with the philosophical foundations of bioethics as an academic discipline and its intellectual heritage, I now want to move from the philosophy department to the clinic and to examine bioethics as a medical discourse which is closely linked to clinical practice. We need to bear in mind, however, that practical or applied bioethics embraces and engages with but sometimes also contests the dominant positions in moral and political philosophy discussed above. Even though bioethics as a discipline and discourse is relatively new—Jonsen situates its emergence in the late 1960s—we can seek its origins in the traditional ethics associated with medicine.[33] Warren Reich, editor of the four-volume *Encyclopedia of Bioethics*, defines bioethics precisely as "the study of the ethical dimensions of medicine and the biological sciences."[34]

It is the need for the regulation of medicine that has prompted the development of this field of study. The Nuremberg Doctors' Trial of

1946–47 provided a strong impetus for putting forward a set of principles that were to guide medical and scientific research at an international level, the Nuremberg Code.[35] The emergence of bioethics is thus clearly associated with the crisis of self-regulation within the medical community and the need to bring in an external regulatory framework. It was only in the second half of the twentieth century that medical ethics became embedded in medicine as a supervisor of its conduct from outside the profession. If one of the points of origin of what became known as bioethics can be located in the Nuremberg Trials, this raises the question of not only whether bioethics functions as a protection against the excesses of the Holocaust but also to what extent it is permanently haunted by its specter. Indeed, if "the wish to control the biological make-up of the population [lies] at the very heart of modernity,"[36] we can wonder about the possible continuity between the biological experiments and the overall "management" of life during the Holocaust, on the one hand, and some of the current practices involved in the positive management of the life and health of populations on the other, shocking and perhaps even distasteful as such a pairing might initially seem. However, this line of interrogation seems inevitable when we consider the proliferation of neo-eugenic discourses and sociobiological arguments at the beginning of the twenty-first century, in the context of TV makeover shows, debates on immigration, and proposals for ubiquitous genetic testing. I would therefore argue that highlighting Nazi eugenics and the Nuremberg Doctors' Trial as its condition of possibility places bioethics in a broader political and cultural framework as well as signals its inevitable and necessary engagement with issues of race, heredity, the technicization of modernity, and the constitution of the caesura between human and nonhuman. (I will return to this point in chapter 3, when I trace the relationship between bioethics and what Michel Foucault has called "biopolitics," a form of political regime under which citizens' lives and bodies are being permanently regulated and managed.)

As a result of the lessons learned from the Nuremberg Trials, the prevailing agreement in medical ethics until the 1950s was that nontherapeutic research should not be conducted and that any medical research undertaken should benefit the patient. These principles fitted in with the ancient Hippocratic tradition, outlining ideal conduct for a physician.

They were adopted, in different versions and translations, by many medical schools in the Western world.[37] Still, with the shift to technology-based experimentation in the second part of the twentieth century and the emergence of new disciplines, such as neurophysiology, organic chemistry, and molecular biology, whose aims and agendas were still very much in the making, the limitations of the earlier principle regarding the necessary prior certitude about the therapeutic value of research on humans and other live agents became apparent. With this technicization of medicine (or, more broadly, the life sciences), the concept of "risk" has entered the field of bioethics. "Risk" is usually coupled with the notion of "benefit," requiring the calculation of the propitious ratio between the former and the latter. In the era of intense biotechnological research, utilitarian perspectives seem to dominate over deontological or virtue-based bioethical positions as ways of responding to this "risk" factor. There is also another dimension to this technicization of bioethics. As well as referring to technology-driven experiments and innovations in the life sciences, it signifies the increased proceduralism and codification of the field, evidenced in the setting up of biopolicy-making bodies as well as national and local ethics committees and councils, and in the emergence of a consensus regarding the availability of universal applicable procedures that can govern issues of health, illness, death, and "life itself." From the 1970s onward, governments in the United States and Europe started to set up institutions and offices regulating bioethics, thus turning the disciplinary debates into predominantly a legal issue.[38]

The legitimization of modern biopower via instances of "giving permission" to science seems to be one of the principal tasks of bioethics today.[39] Proceduralism and formalism are thus two key facets of the dominant positions in bioethics, whose workings focus on providing practical solutions and specific recommendations to determinable problems and case studies. Tom Beauchamp and James F. Childress's *Principles of Biomedical Ethics*, first published in 1979, has become a template for generations of medical students. It put forward four key principles which set a priori normative guidelines for physicians: autonomy, nonmaleficence, beneficence, and justice. Jonsen describes this state of events rather harshly when he says: "[a]lmost from its birth, bioethics was an ethics of principles, formulated as 'action-guides' and little else."[40] He also points out that it is the principle of "respect for persons" that has become dominant in bioethical theory and practice, an idea that ties in with the concept of personal autonomy. This, in turn, has led to the emergence of the principle of "informed consent"—a belief that a patient is an autonomous rational being that requires full knowledge about her medical condition and the required procedures and that should participate in decisions about her health care. The doctor–patient dyad has thus become a cornerstone of medical ethics. This development, offered as a counterforce to the rampant paternalism of much of the medical profession, is of course to be welcomed, but the philosophical principles it relies on—self-enclosed autonomy of the rational human self, the decision-making process as predominantly moral rather than political and cultural—are not without problems as I argued earlier. It is also worth mentioning that the elevation of the principles of autonomy and respect to fundamental moral principles to be obeyed by clinicians and medical researchers poses a challenge to utilitarian and consequentialist moral theories, in which it is only the consequences of an act that determine whether it is moral or not. With a risk of a certain oversimplification, we can say that "conventional" bioethics today finds itself at a theoretical and practical crossroads between three schools of thought: utilitarianism, deontology (i.e., ethics of norms and rules), and virtue ethics (focusing on secular or religiously driven benevolence and charity), with all three underpinned by various humanist principles and normative assumptions.

Bioethics in the Public Domain

It has to be acknowledged, though, that since the 1970s bioethical thinking has started to incorporate the social dimension, with social scientists, philosophers, theologians, and, most recently, cultural and media theorists entering the arena. Bioethics has thus left the clinic and entered the broader social world. As Reich explains, in the United States of the 1970s "there was a political urgency to many of the biomedical issues: consider the groups who warred with each other over abortion and the use of fetal tissue for research purposes. The media craved the biomedical controversies and federal and state policymakers wanted answers."[41] Yet this popularization of bioethics has not obviated the individualized doctor–patient dyad as its structuring relationship. On the contrary, it can be argued that the expansion of interest in bioethical issues outside the clinic and the hospital is one symptom of the general medicalization of populations in the United States and other Western democracies, with every citizen being simultaneously positioned as a patient. It is also a sign of the working of "the biopolitical regime" in which bodies are being managed and through which prescriptive (and often conservative) ideas regarding health and normalcy are being developed. Health and general well-being are perceived here as a moral issue. Of course, associations between health or its lack and morality go back a long time: we can think here of the perceptions of venereal disease or AIDS as God's punishment for promiscuity. However, what was significant about this recodification of health as a moral issue in the second part of the twentieth century was its secularization. The hegemony of the moralizing discourse of health is still prevalent today and can be evidenced, for example, in the recent moral panic about children's increasing obesity and laziness. Indeed, moral panics are one of the main routes through which bioethical issues enter the public domain. Exacerbated by the media—tabloid newspapers, television talk shows, Internet campaigns—the panics concerning genetically modified foods, the triple measles, mumps, and rubella vaccination, the hospital "superbug," or cloning often foreclose the debate about the role of technology and new media in the changing status and nature of the human by resorting to ready-made moralist positions that are presented as universally binding. Bioethics is not, though, just a matter of "bad translation" from knowledgeable science to prejudiced knee-jerk public reaction, as the very same moral assumptions and cultural prejudices regarding "the human" and human life frequently sustain narratives produced by both scientists and the public.

It can be argued that it is precisely in its role and function as a public discourse, one that arises as much out of the concerns, anxieties, and passions of the public as it does out of the disciplinary preoccupations of academicians, that bioethics' vitality and significance can truly unfold. As Jonsen maintains, "The public discourse provides the subject matter for the discipline of bioethics: while we often point to the new science and technology as the cause of bioethics, it is actually the discourse about the uses of science and technology—the differing views and values about human life that inform individual and social judgment about those innovations—that gives rise to bioethics."[42] The recognition that "we are all bioethics experts now," with which I started this chapter, can perhaps be adopted as a call for opening up the narrow professionalism of bioethical discourses and forms of knowledge, even if not for doing away with "experts" altogether.[43] Indeed, Andy Miah suggests that the involvement of the public in bioethical debate can have a number of positive spin-offs: crucially, it can "assist the development of the public understanding of science."[44] He also argues that the subject matter of human genetics and the intimate environment of computer-mediated communications offer a context where the aspiration of the "public understanding of science" agenda can be successfully accomplished. It seems to me that participation in a discourse on the ethics of science and life itself can take the debate beyond its "moral panic" aspect and reposition both science and bioethics as participative, democratic practices that affect, and are affected by, a wide range of social agents.

Alternatives in Thinking about Bioethics

The limitations of the traditional bioethical theories and positions have thus already been recognized by many: we can think here not only of the ongoing critique of systematic philosophy which raises questions for principle-based bioethics but also of the work undertaken by the medical community itself as well as many cultural theorists and feminist philosophers, who have all commented on the gap between abstract, disembodied moral theories and real-life cases. Over the last decade or so, calls for more embedded, less procedural bioethical models that some have described as "postconventional" or "postmodern" have become much more vocal.

The U.S.-based writer Carl Elliott, author of *A Philosophical Disease: Bioethics, Culture and Identity*, describes his work on ethics precisely as "postmodern philosophy": he posits it as postsystemic but also *practical*.[45] Elliott takes the latter attribute extremely seriously and keeps insisting on the "usefulness" of bioethics. His theoretical framework is loosely based on the work of late Wittgenstein, but Elliott has developed a quasi-novelistic method and style of his own—incorporating many anecdotes, venturing into different disciplines, and borrowing equally comfortably from philosophy and cultural studies. His "general antitheory of bioethics" departs from systemic moral theories worked out in advance: instead, he proposes a pragmatic, common-sense evaluation of ethical problems in their broad contexts. Elliott's main issue with traditional bioethics concerns its two aspects: (1) its somewhat oppressive normativity, which is rooted in an "aspiration towards dispassionate objectivity" and detachment[46] and (2) its economic investments and interests. It is the latter aspect of his critique that makes Elliott's work particularly daring. He alerts his readers to the perhaps obvious but often occluded fact that in late capitalist societies medicine is very much part of technoscience and that bioethical debates are implicated in complex networks of political and economic influences—coming from governments, multinational corporations, research institutes, and so on. "And medical ethicists are the practice's maintenance engineers, medicine's moralizing scut monkeys, who outline and argue for certain limitations on medical practice," he adds, not without sarcasm.[47] With this, Elliott raises the important question of how the exchange of money in medicine alters the nature of medical consultation, both on the part of clinicians, who deal directly with patients, and on the part of bioethicists, who provide advice to medical councils and the biotech industry. Elliott is not politically naive: he does not recommend a total severance from the dominant networks of capital (were such a thing even possible). However, he does insist on the need for a greater accountability of bioethics experts, medical doctors, and researchers, as well as a more transparent and public reflection on the vested interests of those who are engaged in dispensing medical advice.

The political economy of life itself, and the way life is inscribed, or even produced, in the network of power structures and power flows dominated by the biotech industry, the clinic, and the academy, is perhaps one of the key political issues today. While Elliott's suspicion regarding bioethicists' too close an alliance with commerce is fully justified, it is worth lending an ear to critiques that raise questions about the feasibility of a radical separation between gift and commodity in the economy of (bio)capitalism. Catherine Waldby and Robert Mitchell, authors of

Tissue Economies: Blood, Organs, and Cell Lines in Late Capitalism, explore the limitations of the "gift culture" based on the "sharing of vitality" in an era when different forms of bodily tissue—blood, eggs and sperm, stem cells—are already positioned within complex global networks of production and exchange and are subject to hybrid "micro-economic arrangements."[48] It can be argued that the heavy mediation of tissues by biotechnological processes and institutions awards them a double status of "natural body parts" *and* "industrial products." (Blood, for example, is rarely transfused as "whole" but rather as fractions, "subsets of blood proteins tailored to suit the clinical needs of the particular patient's condition.")[49] Waldby and Mitchell are no apologists for neoliberal capitalism which commodifies life under the blanket promise of general well-being, but they do acknowledge that the discrete ideas of "gift" and "commodity" are "inadequate to conceptualize their *technicity*, and the ways this technicity mediates the values and relations associated with particular kinds of tissues."[50] If life is already technical, and if any tissue donation immediately enters a set of industrial and technical relations, then what we need is a bioethics that recognizes this technicity and relationality as a condition of bodies' existence in the world, not an unwanted element which has to be eliminated at all costs. Even though the authors of *Tissue Economies* do not provide any specific bioethical pointers, they do encourage us to shift the parameters of the debate on bioethics and life beyond the outdated model of the skin-bound sovereign self and into the complex network of bodily connections, affective and economic investments, technological mediations, and political interests.

A number of other thinkers who derive their work from "postmodern" and "poststructuralist" perspectives have been active in outlining alternatives for less singular, more networked bioethical frameworks. "The body"—as both an unstable locus of subjectivity and an object of technological mediation—usually functions as a key nodal point in such alternatives. In the introduction to the collection of essays *Ethics of the Body: Postconventional Challenges*, which she coedited with Roxanne Mykytiuk, Margrit Shildrick charges traditional bioethics with relying on "models of moral evaluation that derive from a belief in fixed and normative templates as adequate for all new knowledge" and thus effectively duplicating "the master discourse and maintain[ing] the split between a secure sense of the transcendent self as moral agent, and a more or less unruly body that must be subjected to its dictates."[51] She also accuses it of being, both figuratively and literally, "out of touch." Drawing on feminist theory, poststructuralism, and phenomenology, she calls for a bioethics of uncertain responses, "radically transformed by the capacities of bioscience to vary and extend the hitherto limited things of which bodies seem capable."[52] Shildrick's nonnormative ethics for the biotechnological era thus adopts a much more fluid and unstable model of selfhood and embodiment. Embodiment here is not something added that should "also" be taken into account in ethical deliberations: it is a condition of being a self.

Shildrick has already taken some steps toward outlining such a bioethics of an embodied self in her earlier work. In her 1997 volume, *Leaky Bodies and Boundaries*, she makes an astute observation that "the body is curiously absent to us during health, and it is only in sickness that it makes itself fully felt, and then as that which unsettles the sense of self."[53] Calling instead for a positive recognition of the body as a "lived presence," she also acknowledges the instability and internal difference of this bodily presence, which carries markers of difference within it. The body in Shildrick's theory is thus always unstable, but it is also materialized, gendered, and, to use her own term, "leaky." Shildrick is particularly concerned about the normalizing influence of mainstream medical and bioethical practices and the consequences of the differentiations legitimated by these practices for feminist politics. She argues: "The transgressive excess that health care attempts to counter is not peculiar to marginal bodies, but is an integral possibility of all bodies."[54] Set against the traditional medical concerns over the preservation of the material boundaries of the body and the exclusion of otherness—in the form of disease, disability, or virus—from it, Shildrick's bioethics instead recognizes this otherness (of which the body's "leaky status" is just one manifestation) as an intrinsic condition of our being in the world. This being in the world is also always already a way of "being-related"—to other bodies, or, more broadly, other materialities. The ethical moment consists for her in "radical openness to the multiple possibilities of becoming." This she qualifies by explaining that "we should position ourselves among others, claiming no special authority, but without eschewing responsibility either."[55] But it is only through singular situations and cases—the example she studies in detail concerns the use of new reproductive technologies by lesbian couples—that the meaning and moral significance of each particular relation between bodies, knowledges, and values can be worked out, ethically and responsibly.

Some of the most challenging interventions into conventional bioethics have recently come from theoretical positions influenced by the work of Gilles Deleuze, especially by his interpretation of Spinoza's notion of ethics—the echoes of which can be heard in Shildrick's work discussed above. This theoretical development can perhaps be explained by Deleuze's explicit departure from the humanist and organicist vocabulary as well as his putting forward of the notion of the "virtual" as always already "real," and as expanding the scope of potentialities beyond what the human has already imagined or predicted. For Deleuze, ethics is "an *ethology* which, with regard to men and animals, in each case only considers their capacity for being affected."[56] Situated beyond the paradigm of good and evil, it does not seek to reaffirm any fixed values. Instead, ethics refers to the "qualitative difference of modes of existence (good–bad)," the evaluation of which is material or, we could even say, pragmatic, rather than transcendental.[57] The ethical injunction for Deleuze lies in going along with this nonhuman flow of life and expanding life to its fullest potential, beyond the already imagined possibilities. It is "the becoming body" rather than the fixed human subject that is the focus of his ethics. We can see here how this notion of the "becoming body" which is always already machinic, and which is implicated in the ongoing process of life over which the human has no absolute control because she or he is also part of it, has become attractive to researchers of new technologies and new media, especially those connected with biotechnology. In biotechnological processes, living and nonliving elements exist in intimate couplings and associations which imply a design that disallows both the notion of life as something entering a machine in order to animate it and the notion of technology as something added postfactum to the original living entity.[58] What emerges instead is a much more enmeshed model of relations between living and nonliving forms which always already bear a technological inscription.

Claire Colebrook explains that Deleuze uses the concept of the machine to rethink ethics from its "reactive forms" (as a reaction to a pregiven unity that humans as goal-oriented beings envisage) to its "active" status, beyond "an intent, identity or end."[59] With this notion Deleuze describes "a production that is immanent: not the production of something by someone—but production for the sake of production itself, an ungrounded time and becoming."[60] He therefore allows us to theorize life as a technical process, and an ongoing one at that. If life, including human life, is undergoing a constant process of change, then working toward developing new, interesting ways of being and becoming becomes a foundational imperative of Deleuze's (bio)ethics. We could perhaps describe it as a "soft" or minimal imperative, the force of which Colebrook justifies as follows: "A maximised becoming is a commitment to univocity, affirming all those differences and creations which traverse us, including the genetic, historical and affective investments that have constituted us but do not define us once and for all."[61]

The injunction to affirm and work with the differences that constitute our bodies and minds prepares the ground for a different model of bioethics, one in which the distinction between self and other, between inner dynamics and external influence, is not so clear-cut. Drawing on Deleuze's notion of ethics, Adrian Mackenzie admonishes the more conventional bioethical positions for overlooking the ways in which the body contests the prerogatives of consciousness and for normalizing differences by seeing them as secondary and external to the moral subject, not constitutive of it. Mackenzie proposes instead a bioethics that maintains an ethos of "of embodied differences, of the character and habits of individual bodies" and that recognizes the active role of technoscientific intervention in both producing and effacing these differences.[62] In a similar vein, Eugene Thacker turns to Deleuze in the Conclusion to his book, *Biomedia*, in order to develop a bio-ethics—a term he hyphenates in order to foreground its relation to, and rooting in, the body—which would "take . . . up the potential implications of design relationships between bodies and technologies."[63] Driven by Spinoza's question, "What can a body do?" and by Deleuze's rereading of Spinoza, Thacker's bio-ethics is also an "ethics of transformation," with design—of body, matter, and "life itself"—being seen as a bioethical endeavor.[64] This involves, as Thacker explains, a shift toward perceiving bodies as relations between different entities affecting one another, rather than discrete entities. Ethics functions here as an inquiry into the constitution and meaning of the human. This kind of ethics is, in Thacker's words, "nonhuman," even if it does place human concerns as central to itself, precisely because "it does not prescribe human interests, as if the human could be effectively separated from a milieu in which specific embodied humans exist."[65] The recognition of the relationality of the human to other living and nonliving entities radically alters the way we can think about both moral agency and technological influence. It also offers a way out of the moralism of many forms of traditional bioethics which declare in advance that certain procedures and interventions are "good" and "bad" and then apply this checklist to all cases, including future ones.

Spinoza's question "What can a body do?" finds one of its most enthusiastic respondents in the work of the feminist philosopher Rosi Braidotti. (Her answer, as we will see below, is a firm "A lot.") With some help from Deleuze and a number of other poststructuralist thinkers, Braidotti develops a materialist, nomadic philosophy of becoming as an alternative to the simplistic models of biotechnological evolution promoted by "hyper-colonialist capitalism."[66] Siding decisively with the technological forces, she nevertheless opposes "the liberal individualistic appropriation of their potential."[67] As is the case with the other theorists of "alternative bioethics" discussed above, embodiment and embeddedness in the material world are crucial for Braidotti's ethics. She lists the placenta, the parasite, the cloned animal, and the leaping gene as figurations which can be seen as "steps towards a nonlinear rendition of the subject in its deep structures"[68] and as a physical and conceptual opening of the subject to relationality, the self–other interaction, which, in the area of biotechnology, is not just an ethical injunction but also a lived reality. It can be argued that how we live through this hybrid human–nonhuman relationality is precisely one of the key ethical questions today. Working toward a sustainable future while remaining accountable for the constantly becoming world constitutes the goal of Braidotti's philosophico-political project. She is aware that, to be truly ethical rather than oppressive, this project needs to steer clear of the mastering gestures of many Western attempts, both philosophical and political, to "save the world" as a graspable entity, without lending an ear to specific locations, with their singular narratives and practices. It also must remain open to the liberatory and transformative power inherent in, but not guaranteed by, technological forces and processes taking place in the world. The impetus for Braidotti's nomadic ethics of transformation stems from *zoē*: a source of nonhuman, "raw" vitality which she reads—contra Agamben[69]—as a generative force driving both human and nonhuman life.

The Difference of Difference: A Bioethics (Yet) Otherwise

The new way of thinking about life, the body, the human, the animal, and the machine through Deleuzian philosophy is an important step in developing a different framework for bioethics. Pointing to the productivity of technology which is never entirely separate from the human but which rather coconstitutes the world, thinkers such as Thacker, Mackenzie, and Braidotti have made a serious effort in attempting to dethrone human-centered moral philosophy from its position of an arbitrator over the value of life. Positing nondialectical difference as crucial to the possibility of having an ethical future and enacting a political transformation, they have refocused the ethical debate on the question of the body and the processes of embodiment and also have placed ethical issues in a wider context of biocapitalism and globalization.

And yet, as I said earlier, even though when it comes to its intellectual and affective investments my book sits close to the work of many theorists inspired by Deleuze, the way it approaches "difference" is somewhat different. Rather than following Deleuze's understanding of difference as a "power within," an immanent force that ceaselessly produces new forms, it understands the source of an ethical injunction as coming from what Emmanuel Levinas has called "otherwise than being": a place of absolute alterity that cannot be subsumed by the conceptual categories at our disposal. The echoes of Levinas's "otherwise than being" resonate in Derrida's notion of "différance," a spatiotemporal difference which is both the condition of the possibility of meaning and the condition of its impossibility. This place of absolute alterity is thus transcendent—but only in a formal sense: there is no God or other concrete being that legislates this ethical injunction to respond to the alterity of the other. It is the assumption itself—philosophical but also, we may risk saying, experiential—of there being an alterity that exceeds the conceptual grasp, and the very being, of what we understand as the self, that creates the framework for an ethical encounter and ethical event. The difference between these two philosophical positions—one of immanence as encapsulated by the work of Lucretius, Spinoza, Hume, Bergson, Nietzsche, Foucault, and Deleuze, and one of transcendence as developed by philosophers such as Hegel, Husserl, Kant, Heidegger, Levinas, and Derrida—has been articulated most cogently by Daniel W. Smith in his article "Deleuze and Derrida, Immanence and Transcendence: Two Directions in Recent French Thought." Smith states that while Deleuze tries to expunge from Being all remnants of transcendence, Derrida seeks to trace the eruptions and movements of transcendence *within* Being. Deleuze's difference is defined "in terms of a genetic principle of difference,"[70] as differentiation within life itself, and is strictly linked to becoming.[71] This is why he can reconcile this idea of difference and differentiation-from-within with his concept of univocity (an idea, as Smith explains, which Deleuze borrows from Duns Scotus), where a certain link or kinship is posited between all forms of life, both real and virtual ones. For Derrida, in turn, difference always already involves a "cut," an impossibility of the ultimate connection. It is precisely in differentiation that cannot be sublated or tied in to any entity that an ethical demand and ethical impression on the self take place.

Thus, even if we agree with Paul Patton and John Protevi when they say that "Derrida and Deleuze share an ethico-political conception of philosophy as oriented towards the possibility of change,"[72] we may also perhaps say that Deleuze is more interested in the flow of life, while Derrida pays more attention to a cut or interruption to this flow. It is precisely this cut, a differentiation within the flow of life that cannot be subsumed by this life because it comes from (formal, not theological) "elsewhere," that will constitute a pivotal point of entry for my own attempt to rethink bioethics in this book. Now, the difference between the two positions or indeed traditions—because, as I explained above, we are not just setting Deleuze against Derrida here but rather considering two parallel, even if not entirely separate, philosophical lines of thought that can be aligned under the headings of "immanence" and "transcendence"—is not purely academic. It reflects a broader set of concerns that are important for my search for bioethical alternatives: the role of negativity in ethics, the position of desire in its relation to both productivity and lack, the graspability (or not) of otherness that drives this ethics, the understanding of power and its relation to oppression and liberation. My own position springs from a certain suspicion toward Deleuze's "joyful affirmation" and a concern over what we could describe, perhaps too briskly, as "theoretical neuroticism" that seems to drive it. This is not to say that a bioethics I want to envisage, with some help from a number of thinkers inspired by the philosophy of nonreducible alterity, will be "negative" but rather that it will recognize the place of negativity, lack, and what I have described above as "the cut" in the flow of life as constitutive to it. The latter will also allow us to consider the differential relation between humans, animals, and machines in this ethics while also attending to the nature of this difference, always in a singular way—as a question that needs to be asked over and over again. It will thus deal with issues of limitations, restrictions, and fractures as much as it will with transformations and potentialities. Last but not least, the bioethics rooted in the position of infinite alterity will suspend us between body and language, not in the sense that the materiality of the body will be reduced to language but rather that, to be properly understood, in both its multiplicity and its lack, and to be able to meaningfully act and enact, the body will require linguistic articulation. As Judith Butler puts it in her rejoinder to Rosi Braidotti's work (with which she does recognize numerous philosophical and political affinities), "The body is that upon which language falters, and the body carries its own signs, its own signifiers, in ways that remain largely unconscious."[73] So we are back here with lack and "the cut" as both a limitation that has to be worked through and a condition of possibility.

However, let me repeat it again: my position is not anti-Deleuzian, rather "alongside Deleuzian," because I share many of the affective investments that structure Deleuze's philosophy and that of numerous other thinkers who draw on his work—an investment in the idea of the transformation of life, in envisaging better, more free ways of living, or in attending to the relationality between different life forms. I just find different points of entry into addressing these issues. Rosalyn Diprose, who has made a serious attempt to think about bioethics from the position of alterity, seems to be thinking in a similar vein when she argues that ethics predicated on the infinite alterity of the other ensures the possibility of a real transformation of both the self and the world. She writes in *Corporeal Generosity* that "it is the other's alterity that makes me think, rather than ideas I live from and that seem to make me what I am. It is this alterity that provokes any gesture of expression, is necessary for its production, and is not subsumed by the incarnate thinking that results."[74] In her earlier book, *The Bodies of Women: Ethics, Embodiment and Sexual Difference*, Diprose argues that it "is about being positioned by, and taking a position in relation to, others"[75] and that both our "being" and the "world" are constituted by the "in" that connects them. In other words, the relationship between embodied place and the social world is constitutive for her. However, Diprose also insists that ethics must pay attention to different ways in which these relations between various beings are established. "[I]f ethics is about taking a position in relation to others then it is also about the constitution of identity and difference," she adds.[76] These instances of the constitution of identity and difference also become what I call "temporary points of stabilization" in the process of change, in the becoming of matter (which Deleuze refers to poetically as "becoming-woman," "becoming-animal," and "becoming-machine"). The study of these points of stabilization becomes an important task for bioethics today, but any such study needs to be undertaken from the embedded position of lived experience, of already being embedded, immersed, and connected—and being able to grasp the meaning of this connectedness.

Even though the inspiration for the bioethics I want to outline throughout this book comes from the tradition of the philosophy of difference, this tradition needs a clearly articulated supplement to which I have been referring throughout this chapter: that of a more engaged relationship with technicity. Such a supplement will allow us not only to overcome the humanism of Levinas[77] but also to take the bioethics of life beyond the context of the clinic and the lab and into the multiple territories of everyday life where new technologies and new media are constantly engaged in redefining the constitution of the human, of human and nonhuman life: virtual gaming environments, social networking portals, televised aesthetic surgery, biotechnological experimentation and commercialization.

As the reader will have hopefully realized by now, the role of the critical overview of what I tentatively describe as "traditional bioethics" presented in this chapter has not therefore been merely descriptive. By probing into the intellectual tradition, philosophical assumptions, and disciplinary boundaries that demarcate dominant positions in bioethical thought, we have also begun to see emerge a different framework for thinking about bioethics in the age of new media: one that does not fully negate the existent framework but that rather emerges at the margins of the dominant system of thought as the very system's not always acknowledged aberration or exception. This bioethical framework, which I will articulate and enact in more detail in the following chapters, will not rely on a priori principles that could then be applied to selected cases, nor will it be involved in a mathematical calculation of goods in order to satisfy the greatest number of desires and preferences. Informed by the need to provide a constant response to the alterity of other beings and life forms, and for a decision, always to be made anew, about what to do, it will nevertheless assume responsibility for the lives and deaths of multiple, human and nonhuman, others, even if the locus of this responsibility, response, and decision will not be clearly located in a bounded, rational, human self. Putting into question the concept of the human, body, and life—and thus also many of the concepts positioned as the human's "other," such as animal and machine—the bioethical enquiry, the way I envisage it, will also involve an examination of the historical formation and ideological structuration of "the human" and the related concept of life. The human will thus not disappear entirely altogether from this enquiry. Instead, this "not yet human, never fully human" will become a strategic point of entry for this nonfoundational, aprincipled bioethics. But it is first of all the relationality of the human—his or her emergence through technology, his or her material coexistence in the sociocultural networks, and his or her kinship with other life forms—rather than the human's positioning as a cognizable, disembodied, separate moral unity that will drive my efforts to think bioethics otherwise. The new bioethical framework will seriously consider the temporary moments of stabilization, with all their accompanying and inevitable violence, as meaningful instances in which "something emerges" and "something happens," but whose fleeting ontology does not arrange itself into a stable image of Being.

It can be argued that the interdisciplinary project of media and cultural studies, which has been actively engaged both in studying "the discourse about the uses of science and technology" and in seriously considering the public discourse as constitutive of what counts as "culture," "morality," and "politics," is rather well predisposed to undertaking a bioethical enquiry in the age of new technologies and new media, and taking it in a radically different direction. Indeed, media and cultural studies, informed by work conducted under the aegis of science and technology studies, anthropology, sociology, and feminist theory, can provide a necessary rejoinder to the classical disciplines—philosophy, theology, law, and medicine—that have shaped the field of bioethics. It is with these concepts as tools in hand—alterity, technology, new media, culture, and embodiment—that I want to move on toward outlining or maybe rather enacting this alternative bioethical framework in the chapters that follow.

2

A Different History of Bioethics: The Cybernetic Connection

The "Bilocated Birth" of Bioethics

Bioethics—by which I mean the interrogation of "ethical issues arising from the biological and medical sciences"[1]—has come to occupy a significant place on the public agenda. As my preface makes clear, many social groups, encouraged and provoked by the media, have been engaged in an ongoing debate over issues concerning our life and health and the medical interventions into both: abortion, doping, cosmetic surgery, or genetic testing. However, it is the transformation of the very notion of life and of the accompanying notion of "the human," as well as the promises and threats to human and animal health posed by science and technology, that have evoked particular hopes and anxieties among the public in Western liberal democracies. As a discipline that combines theoretical insights of moral philosophy with the experience of clinical practice, bioethics has been mobilized with the task of having to arbitrate over life, death, and the nature and role of the human in the age of digital technologies in a number of different forums: in the media, in scientific research committees, in hospitals, and in biotech companies. At the risk of overgeneralization, I want to suggest that its response to this task has so far been rather conservative, in the sense that the foundational humanism of the theories and practices that traditional bioethical discourses have been based on, be it in their religious or secular guises, has remained intact in a great number of recent bioethical debates—and this, in spite of the fact that genetic patenting, cloning, xenotransplantation, cochlear and corneal implants, and organ printing have radically called into question not only humans' ontological status as skin-bound, sovereign beings but also their kinship with, and dependency on, other species and material forms.

The aim of this chapter is to engage with this inherent humanism of bioethics and consider the possibility of thinking bioethics otherwise, beyond the belief in the intrinsic dignity and superior value of the human, and beyond the rules and procedures rooted in this belief. It is also to challenge what we may call a "cognitivist presumption" of humanism, that is, the conviction that the human can be distinguished from other forms of life by the intrinsic "truth" and teleology of his or her being which is to be revealed to him or her, and which he or she can uniquely grasp.[2] What I want to outline in response is a nonsystemic bioethics of relations which does not abdicate its ethical responsibility or its political commitment, although it does question many of the premises of established moral and political theories. Indeed, I want to "put to the test the pretentious belief that only a liberal and humanistic view of the subject can guarantee basic elements of human decency: moral and political agency and ethical probity."[3] However, my project is more genealogical than futurologist. By revisiting the story of bioethics' emergence as a discrete discipline, I want to address the following question: what if the cyborg rather than the human had been adopted as its foundation? Or, to put it another way, what would a bioethics for humans, animals, and machines look like?

The work of invention hardly ever starts in a vacuum. My efforts to envisage a new, alternative bioethics, whereby the human is not being posited as a central value or datum point but is rather considered *in-relation-to* and *in-difference-with* other life forms, will benefit from revisiting what we may call bioethics' "foundational moment." This genealogical excursion will be undertaken with a view to tracing such an alternative already present at the very inception of the discipline of bioethics. In what follows, I thus want to lead the reader through an early but now forgotten conceptualization of bioethics, one in which the relationship between the human, technology, and the world was understood in terms of a dynamic system rather than discrete entities. The term "bioethics" was first suggested by Van Rensselaer Potter, an oncologist working at the University of Wisconsin, who drew on biology, evolution theory, and cybernetics to develop his proposal for an ethics of obligation to the biosphere as a whole. However, his is not the only contribution to the invention of bioethics. The historian and theorist of bioethics Warren Reich traces the emergence of bioethics as both a concept and a discipline in two articles published, respectively, in 1994 and 1995 in the *Kennedy Institute of Ethics Journal*. He claims that in 1970–71 bioethics actually "experienced a bilocated birth," with Potter and André Hellegers, a Dutch obstetrician and fetal physiologist at Georgetown University who became one of the founders of the field-defining Kennedy Institute of Ethics, developing their theories of bioethics simultaneously.[4] Hellegers himself was in fact closer to Potter's global approach to bioethics, but the other scholars working at Georgetown adopted a much narrower definition of the term. Rather than encompassing the Potterian sense of obligation to the biosphere as a whole, for Georgetown scholars bioethics effectively became "a revitalized study of medical ethics," rooted in applied normative ethics and dealing with "concrete medical dilemmas."[5] Reich points out that "at the outset . . . Potter's use of the term 'bioethics' was marginalized, whereas the Hellegers/Georgetown biomedical connotation of the word came to dominate the emerging field of bioethics in academic circles and in the mind of the public."[6]

Reich's narrative about the "bilocated birth" of bioethics is fascinating in its insight into the institutional dimensions of the legislation of this particular academic discipline and the procedures and exclusions involved in its founding. Following in Reich's footsteps, I would like to retrace the story of bioethics' less popular twin—the one fathered by Potter, focused on the biosphere as a whole, and concerning both medical and environmental issues. The impetus behind this genealogical study of bioethics, as explained above, lies in my attempt to identify the already existent possibilities for thinking bioethics otherwise which are inherent to the discipline itself. Even though, as we shall see below, Potter's bioethics was not necessarily free from the humanist assumptions shared by the Hellegers school, it nevertheless outlined a networked model of relations between the human, technology, and the environment in which the human was seen as part of a larger system. This forgotten model, which, for the sake of simplicity, we can call "bioethics as cybernetics," will provide a useful directive for my own experiment in thinking bioethics otherwise (even if I will ultimately part ways with the cybernetic legacy).

Cybernetics: Bridging the Mechanical and the Organic

Cybernetics played a significant role in the humanities and social sciences in the second half of the twentieth century. Indeed, it is via an encounter with cybernetics and systems theory that the intrinsic and foundational humanism of the humanities was initially challenged, while also allowing for a (not always easy) conversation between different disciplines.[7] The cybernetic model of the world outlined a new epistemology based on "the interaction between variables within a system rather than the domination of the whole by one of its parts."[8] It also put forward the assumption—echoed in the humanities and social sciences' subsequent turn to "the cyborg" as a new model of subjectivity in the era of technoscience[9]—that "there is no necessary contradiction between the organic and the technological."[10] As Katherine Hayles explains, "Cybernetics was born when nineteenth-century control theory joined with the nascent theory of information. Coined from the Greek word for 'steersman,' cybernetics signaled that three powerful actors—information, control, and communication—were now operating jointly to bring about an unprecedented synthesis of the organic and the mechanical."[11] Ambitious in its almost holistic interdisciplinarity, cybernetics was conceived as a science that was supposed to provide an explanatory framework which would jointly consider humans, animals, and machines as information processors.[12] The merging of ideas from biology and mechanics led to the inclusion of both living organisms and mechanical ensembles within the category of "the system" and to the supposition of structural continuity between humans, animals, and machines.

Even though the latter notion proved extremely fruitful for humanities and social sciences scholars interested in exploring less bounded and more relational models of human subjectivity, it is hard not to notice the political conservatism implied by the idea of the system's homeostasis, that is, the belief that all systems maintained their internal stability, no matter what impulses they received from the environment. However, things changed during the so-called "second wave of cybernetics," a period between 1960 and 1980 informed by the work of Humberto Maturana and Francisco Varela. Maturana and Varela saw the world as a set of informationally closed systems which nevertheless remained open to the environment on the level of their material structure. As Hayles explains,

> Organisms respond to their environment in ways determined by their internal self-organization. Their one and only goal is continually to produce and reproduce the organization that defines them as systems. Hence, they not only are self-organizing but also are autopoietic, or self-making. . . . In a sense, autopoiesis turns the cybernetic paradigm inside out. Its central premise—that systems are informationally closed—radically alters the idea of the informational feedback loop, for the loop no longer functions to connect a system to its environment. In the autopoietic view, no information crosses the boundary separating the system from its environment. We do not see a world "out there" that exists apart from us. Rather, we see only what our systemic organization allows us to see. The environment merely *triggers* changes determined by the system's own structural properties. Thus the center of interest for autopoiesis shifts from the cybernetics of the observed system to the cybernetics of the observer. Autopoiesis also changes the explanation of what circulates through the system to make it work as a system. The emphasis now is on the mutually constitutive interactions between the components of a system rather than on message, signal, or information.[13]

The most radical shift here concerned the role of the observer, who was now seen as part of the system: she was embodied and embedded within the system, rather than maintaining an immaterial God's eye view over it. The consequences of this shift were not just epistemological but also ontological, as the observer was in fact creating a world through the acts of cognition and linguistic articulation she was engaged in. What became particularly interesting for humanities-trained media scholars about the cybernetic propositions was cybernetics' radical critique of both realist epistemology and scientific objectivity, an approach that shares many affinities with the tradition of poststructuralism in the humanities.

Cary Wolfe's work in particular provides a very persuasive translation of Maturana and Varela's ideas on systems and self-creation into the humanities. According to Wolfe, "systems theory promises a much more powerful and coherent way to describe the complex, intermeshed networks of relations between systems and their specific environments of whatever type, be they human, animal, ecological, technological, or (as is increasingly the case) all of these."[14] It thus offers us a possibility of theorizing a politics and an ethics that do not have the human as their datum point around which everything revolves but which itself remains free from theorization. In the words of Maturana and Varela, "everything said is said by someone,"[15] which means that when the observer—herself a smaller discrete system within a larger system—makes statements about reality, these statements are always produced *from within* the system. In other words, reality comes into existence for us "only through interactive processes determined solely by the organism's own organization."[16] Cognition and articulation are at the same time acts of "bringing forth a world." The totality we arrive at is thus always already "our" totality and hence is inevitably partial, fragmentary. Wolfe explains this paradox in the following way: "although second-order systems theory does make a claim to universal descriptive veracity, that claim is mounted upon its ability to theorize the *inability* to see the social or natural system as a totality from any particular observer's point of view."[17] We can see here similarities between cybernetics and poststructuralism, especially deconstruction and its elaboration of the concept of performativity,[18] theories that have had a significant impact on rethinking politics and ethics beyond the certitudes of linear historical trajectories and beyond tradition-sanctioned moralism.

Significantly, in their introduction to the special issue of the journal *Cultural Critique* on "The Politics of Systems and Environments," William Rasch and Cary Wolfe go so far as to identify in systems theory a promise for the Left that has lost its political confidence and effectivity because its theoretical ground has been exhausted.[19] They associate the latter loss with the realization that representation-based politics, whereby its subjects imagine collectively a world to be and then strive for its emergence, and which positions the "working class" as representative of the democratic struggle, does not work anymore. And yet the authors seem to turn a blind eye to the fact that the affective investment in the structural idea of the system that characterized traditional Left politics is carried through in this move to systems theory, even if the content of this investment has changed—from "revolution" to "self-creation."[20] Disappointed with both the unfulfilled promises of totalizing Left politics and the open-endedness of deconstruction, Rasch and Wolfe justify their turn to systems theory by claiming that, like pragmatism and deconstruction, it "insists on the social and historical contingency of all knowledge and on the impossibility of thinking unity, origin, or ground." However, they also add that "[u]nlike deconstruction," systems theory "attempts to provide a coherent means of describing all systems—inorganic, psychic, social."[21] In this way, they reveal their aforementioned affective investment in the ideas of totality and closure, the ideas whose loss in traditional Left politics they seemingly feel melancholic about. It is worth emphasizing again that we are dealing here with a "partial" or "strategic" totality, because the system, after Maturana and Varela, only remains closed on the level of organization, not on the level of structure. The latter means that the system preserves its dynamics, that is, the relationship between its parts or particles, but not necessarily all the actual particles that constitute it. (To take an example of a tornado, different particles get swept into the tornado and then dropped from it, but the dynamic relationship between the particles which at a given moment are constituting the tornado must be maintained.) However, this distinction between structure and organization is not entirely stable, especially when we conceive of systems in genetic terms. It thus seems that Rasch and Wolfe *prefer* to see the world as systemic, even if they present this preference as a new theoretical perspective for the humanities and not an ontological truth.

Suspicions arise also as to the extent to which the implicit conservatism of the "first wave of cybernetics," which foregrounded the importance of maintaining the system's status quo, actually disappears in its "second wave," in spite of the fact that its key concept of autopoiesis allows for transformation and emergence *within* the system—be it a cell, a tornado, a human being, or a polity. When a rather messy element such as the observer is incorporated into the system, the emphasis is still on the system's *functioning*, not on its potential disentanglement. Thus, any too radical a change is warded off by an appeal to the system's "natural" closure. One may wonder, for example, how this "new" model of political theorization would work if applied to instances such as suicide bombing or death by drug abuse, in which the system's organization undergoes a transformation on a number of levels. It is also worth pondering over what would distinguish a cybernetic response from the standard "moral" condemnation of such practices. The pronouncement by Norbert Wiener, one of the founders of cybernetics, that "Entropic decay is evil; entropy is morally negative"[22]—where entropy is the opposite of "organization"—only adds to this doubt.[23]

We can therefore ask why a humanities scholar would prefer to look for such partial totality of the system at all, why systems theory is a desirable way of thinking about the world, its politics, and ethics. Of course, we can accept a *strategic* need for identifying such "total but partial" systems—for example, if one is a systems engineer trying to identify bugs in a particular network of computers or a medical doctor administering an appropriate amount of insulin to a diabetic patient. However, even these instances of "strategic" recourse to autopoietic, that is, "open but closed," systems are not culturally, politically, or ethically neutral as they rely on a prior valorization of certain states of being, or the system's organization, as intrinsically "better." This is not to say that we cannot evaluate democracy, addiction-free life, or organic food as more desirable in a particular set of circumstances, in a given historical and political situation we find ourselves in—as long as we reflect on the conditions of emergence of this valorization. Problems arise when these values are presented as inherent and natural, as always ensuring the most optimum functioning of "the system" (cell, human body, Iraq), no matter what the cultural context of a given instance is.

Indeed, "culture" seems to have been one of cybernetics' blind spots. By saying this I am not advocating replacing the science of cybernetics with "cultural relativism" or "constructionism"; I am only foregrounding the fact that no decision we make can be made on universal grounds, as we always operate in a culturally specific situation. For example, ideas and entities such as "organism," "gene," "woman," "animal," and "machine" always bear cultural inscriptions—which does not mean that they are "just constructed," or that they are mere figments of our imagination. It is therefore in culturally specific situations that our narratives about the world and its organization emerge. Narratives, as Katherine Hayles points out, can nevertheless be dangerous "for someone who wants to construct a system,"[24] precisely because they prompt us to reveal the system's very foundations, that is, its conditions of possibility and impossibility. And yet many constructors (i.e., observers) of systems seem to forget about their own affective investments in the drawing of a system. For example, in a very lucid book that attempts to take systems theory into the twenty-first century, titled *Network Culture: Politics for the Information Age*, Tiziana Terranova mobilizes systems theory for her own neo-Marxist political project. Her aim is to redeploy the control mechanisms identified by cybernetics in order to get the system to do what a Marxist would want it to do. While I broadly share Terranova's commitment to certain forms of leftist politics, I have problems with the strategic totality, manifested in an attempt to control the system for the higher aim that she, together with those who share her political allegiance, defines for it. What she thus ends up with is politics as a technocratic application, underpinned by a moralism for which no account is given. The sliding between types of systems, between bug colonies and Internet users (justified by turning to Deleuze and Guattari, Italian autonomists, and Hardt and Negri), creates another form of biologico-political totality, whereby the theorist's desire for control, and the activist's decision as to which nodes actually form the multitude and which are excluded from it, are obfuscated by the rhetoric of fluidity, emergence, and self-organization which hints at the naturalness—or at least control-free spontaneity—of the processes of political change. Even if Terranova claims that such systems are not easily controllable and predictable, as "[a] multitude of simple bodies in an open system is by definition acentered and leaderless,"[25] her argument focuses on systems' *working*, not on their condition of emergence, the observer's cognitive and articulatory labor, or the processes of inclusion and exclusion which are inevitably involved in the drawing of the system's boundaries. This is not to dismiss the otherwise brilliant analysis Terranova provides. It is only to call for a narrative supplement to systems theory, a narrative that would allow us to reconceptualize the ontological and political status of "the multitude" as a new form of political organization and to consider who does not belong to it—and why.[26]

However, let us not throw out the cybernetic baby with the systemic bathwater (or flow) just yet. Hayles points out that "systems theory needs narrative as a supplement just as much, perhaps, as narrative needs at least an implicit system to generate itself."[27] Any academic work consisting in putting forward an argument or a theory inheres at least a partial investment in the systematicity of language and of the disciplines within which one is working. I am thus wary of setting the "culturalism" of the humanities against the "technicism" of cybernetics. Following Gilbert Simondon and Bernard Stiegler, I want to suggest instead that we go beyond the adverse conceptualization of "culture" as a system of defense of humanity against technics, and pay attention to what Simondon defines as the dynamic of technical objects, whereby the human is no longer seen as an intentional actor but rather as the operator of this dynamic. Culture therefore needs to be "adjusted" to technics, which means not only understanding the technicity of contemporary machines and information systems but also acknowledging "technical dynamics" as the condition and foundation of culture, not its opponent. From this point of view, technological evolution, or, more broadly, systemic change no longer has an anthropological source, even if the role and presence of the human is still maintained in this nonhumanist theory of culture and/as technics.[28] The importance of Maturana and Varela's insight that systemic organization is key to our perception and comprehension of the world, and that we make (sense of) the world by arranging it into systems, becomes evident here. However, we should perhaps also borrow from Derrida's deconstructive thought an injunction to trace systems' foundations, their closures and openings as well as the exclusionary mechanisms that are always at work in setting up a system.[29]

Although we should remain aware of the narratological character of scientific discourse (which, I repeat, does not amount to saying that science is just "made up"), we should also acknowledge the significance of the cybernetic articulation of living beings, including humans, in non-anthropocentric terms, as well as of its foregrounding of the relationality and circularity of different living organisms.[30] Cybernetics has put a challenge for the humanities, traditionally centered around the idea of the human as the dominant producer of culture and the world, to imagine different ways of understanding and articulating the world, its signs and traces.[31] A number of philosophers have already taken up this challenge. Stiegler, for example, has drawn on the paleontological theories of André Leroi-Gourhan to argue that the drive toward exteriorization, toward tools, artifice, and language—in other words, toward *tekhnē*—is due to a technical tendency which already exists in the older, zoological dynamic. It is due to this tendency that the (not-yet) human stands up and reaches for what is not in him: and it is through visual and conceptual reflexivity (seeing himself in the blade of the flint, memorizing the use of the tool) that he emerges as always already related to, and connected with, the alterity that is not part of him. "For to make use of his hands, no longer to have paws, is to manipulate—and what hands manipulate are tools and instruments. The hand is the hand only insofar as it allows access to art, to artifice, and to *tekhnē*,"[32] writes Stiegler. The human is thus always already prosthetic; relationality is the condition of his emergence and being in the world. Being in the world amounts to being "in difference," which is also—for Emmanuel Levinas, as much as for Stiegler—being "in time": that is, having an awareness and a (partial) memory of what was before me and an anticipation of what is to come.[33] In this way, Stiegler takes issue with the myth of the originary self-sufficient, total man, living in the state of nature—a myth which is still rather potent in many contemporary articulations of the fears and anxieties concerning technology—as the state of nature stands "precisely [for] the absence of relation."[34] As such, it marks the impossibility of the human (and also of tool use, art, language, and time).

The narrative of systems theory, with its detailed analyses of how particular living organisms function, therefore allows us to conceptualize the closeness of humans, animals, and machines in a radically different way. Drawing on the scientific work in cognitive ethology, field ecology, and cognitive science, which have all demonstrated that the traditional marks of the human, such as reason, language, and tool use, are found beyond the species barrier,[35] cybernetics and systems theory have enabled scholars of human-centered disciplines to rethink the concept of the human as the king of "the chain of being" and to outline a new posthumanist epistemology. Inspired by cybernetic propositions, Van Rensselaer Potter was the first to have taken significant steps toward a posthumanist bioethics.

Potter's Bioethics: A Bridge to the Future

In his 1970 article "Bioethics, the Science of Survival," published in the journal *Perspectives in Biology and Medicine*,[36] Potter put forward a suggestion that ethics should be understood in a broader ecological context and that "[*e*]*thical values* cannot be separated from *biological facts*."[37] Potterian ethics was thus closer to what we call today "environmental ethics," as it examined humans' relationship to land, and the animals and plants that grow upon it, as well as posing questions about the dominant economic paradigm of thinking about land. According to Potter, humans should think of obligations they have toward the ecosystem rather than privileges that they allegedly enjoy. The survival of the whole ecosystem and the improvement of the quality of life would be the test of the new value system proposed by Potter. Bioethics thus became for him a starting point of the science of survival. Exploring both medical and environmental issues, Potter managed to take bioethics from the narrower domains of moral philosophy and clinically oriented practical enquiry and situate it in a wider sociopolitical nexus. And thus his bioethics included "ethical issues in public health, population concerns, genetics, environmental health, reproductive practices and technologies, animal health and welfare, and the like."[38] Potter can therefore be seen as offering an alternative to the medicalized model of ethics developed by the Hellegers school, a model which has become one of the dominant tools in the biopolitics of Western democracies.[39] Potter's importance also lies in his pointing to the connections between different processes and practices, both biological and cultural ones, involved in the management of human life and health. (As an oncologist, he was aware of the links between genetic and environmental factors leading to cancer, agricultural policies regulating the growth of tobacco, and health care and cancer therapy as well as health education.) It is this connectedness between different levels and practices that gave his bioethics a "global" scope.

In *Bioethics: Bridge to the Future*, a book that followed his pioneering 1970 article, Potter conceptualizes bioethics as a bridge between biology, social sciences, and the humanities. "From such a pooling of knowledge and values may come a new kind of scholar or statesman who has mastered what I have referred to as 'bioethics,'"[40] he proclaims. However, he also calls for the integration of the reductionism and mechanism associated with molecular biology—the paradigm he draws on, although not without reservations—with "the holistic principles."[41] Applying the principles of organization and cybernetic theory, he considers the human as "an information-processing, decision-making, cybernetic machine whose value systems are built up by feedback processes from his environment. These feedback processes are embedded in the most primitive forms of life, and they form a continuous spectrum all the way back through prehistory and to times when no life existed."[42] Potter argues that "the essential message is one in which disorder, or randomness, is used to generate novelty, and natural selection then generates order."[43] The turn to cybernetics and organization theory is significant here, as it challenges human-driven and human-centered versions of moral theory. The human is still the principal legislator of the values of the system in which he finds himself, but it is the wider processes taking place between the human and his environment that are crucial to both comprehending the world and ensuring its continued existence. Bioethical agency is partly displaced here from the human to the system—of which the human is part. To use Stiegler's vocabulary, the human is the system's "operator" rather than its "intentional actor."

Potter's proposition that the human organism is a cybernetic system must have been inspired by, among others, the prominent biologist Jacques Monod. Monod viewed the organism as "a self-constructing machine" which "calls for a cybernetic system governing and controlling the chemical activity at numerous points."[44] This position entails a radical revision of an earlier, one-dimensional machinic view of the human, a view implying "the enslavement of the human being by technology and technological determinism."[45] In the cybernetic view, as explained earlier, the human is not seen to be slavishly adapting to the requirements of technology; instead, human adaptation, a positive and necessary characteristic of survival, is a case of the interaction of variables within a system, not a one-dimensional determinism. Significantly, technology is not positioned here as a dangerous "other" which threatens the original purity and innocence of "man." On the contrary, human "origin" and being are understood as always already technological. (As we saw earlier, this idea has been substantially developed in the post-Heideggerian thought of Stiegler.) Cybernetics thus offers a challenge to the Edenic myth of originary nature, which still actively feeds into contemporary moral panics about genetically modified (GM) foods, aesthetic surgery, and cloning. And yet there are also problems with this new, "posthuman" view. As Kathleen Woodward suggests, the proliferation of cybernetic rhetoric in descriptions of the human results from too easy an application of an open metaphor whereby both computers and human beings start to be perceived as two different species of the genus "Information Processing Systems."[46] This mode works by focusing on a few determinants which then get designated as the basic "program" of the human behavior that responds to the "demands of our environment." Woodward thus argues that "[t]he metaphor of the human organism as cybernetic machine"—one that is espoused by Potter and that implies "to a large extent a *critique* of technocratic control"—actually ends up reinforcing it.[47]

Bioethics and Cybernetics: A Bridge Too Far?

Potter's bioethics therefore entails a number of conceptual stumbling blocks. It is based on the presupposed analogy between biology and culture (an analogy which actually precludes the analysis of the two in terms of a dynamic *relation*) and on the belief that the analysis of the processes occurring at the microlevel of the cell can be extrapolated to the higher level of the environment and thus culture—an idea he appropriates, somewhat erroneously, from the cultural anthropologist Clifford Geertz. It is the processes of adaptation and natural selection at both biological and cultural levels that Potter sees as vital for the progress of the world. Progress is measured in terms of the survival and prosperity of the human species in the changing environment, a state of events that should permit "every man to develop to the maximum his inherited talents."[48] Potter asserts: "I believe [most scientists] would agree with John Dewey that progress consists of movement toward a society of free individuals in which all, through their own work, contribute to the liberation and enrichment of society as a whole. I believe that a revitalization of our value system is both necessary and possible."[49] He thus presents us with a curious mix of humanism, individualism, and liberalism, coupled with a belief that the world is a system with built-in mechanism for disorder. Thus, even though the human is not the driving force of this world—because "the system operates as it can, not as we would have it"[50]—it is nevertheless the human (in this case, Potter himself) that defines in advance the world's values. These values are human survival, belief in man's need to realize his full potential, human dignity, happiness, and the prosperity of the human species, although Potter tends to make them sound like scientific facts, that is, products of the optimization and adaptation of the human species, rather than moral injunctions. Hence Potterian bioethics emerges as a result of adjustments after a certain "minimal amount of disorder," which is seen as indispensable for biological evolution and mutation, has been taken care of by the system. Even though creativity and change are regarded as inevitable and even desirable for the survival of the environment, society, and humankind, the minimization of disorder and the restoration of systemic unity are also positioned as both necessary and always already calculated into the system itself.

Bioethics ultimately becomes here an application of the cybernetic principles of feedback and control to the behavior of the system, a comparison of the action taking place in the world against the principles of that world. Potter argues: "Every action is governed by feedback mechanisms in which the action is designed to close the gap between what the human cybernetic machine is doing and what it believes it should do. In other words, feedback mechanisms always involve a reading of an action and a comparison of the reading with a preset standard. This standard may be set by beliefs and tends to shift with time, but the man machine is always trying to close the gap between the actual performance and the standard."[51] The actual process of the working out of standards is not something Potter is concerned about too much, as this will be taken care of via the rational deliberation between experts, who will be focusing on "advanc[ing] the human condition."[52] The orientation and meaning of this "advance" presumably do not need to be investigated further, as its very origins are biological: they lie in the organism's natural affinity for organization and order, acquired in the evolutionary process. If autopoiesis "establishes a sphere of existence for the individual, a location from which the subject can ideally learn to respect the boundaries that define other autopoietic entities like itself,"[53] it is conservative, in that the self remains closed to an alterity that it does not know or recognize, an alterity that can threaten its stability (closure) and destroy it. It is therefore actually rather nonethical.

We could perhaps suggest, then, that it is the mechanical side of bioethics, that is, the unimpeded running of the system, that is of primary concern to Potter. Indeed, evolutionary processes are seen as the source and origin of all values, that is, of our morality, with Potterian bioethics acting as another control mechanism to the self-adjusting system. Culture itself is understood as an effect of an evolution guided by natural selection. And this is partly where the problem with Potterian bioethics lies. If cultural evolution is seen as analogous to biological evolution,[54] any forms of antagonism or rupture—what Potter would term temporary disorder or randomness—need to be overcome while the system is being adjusted to the new circumstances. As, according to Potter, rational human beings, guided by experts, should all agree on the direction in which the world should develop and the minimal values that drive it, he can unproblematically present his ethics as beneficial for the whole society, which he sees as constituted out of "free individuals." "Individualistic aspects" of culture are seen as variations resulting from adaptation, not as structuring elements of a social order. In the last instance, they can be seen as moments of disorder and creativity within the system—but not as conditions of the system's (im)possibility of being. Thus, it is not even the lack of recognition of what is sometimes referred to dismissively as "cultural difference" that is problematic about Potter. It is the elimination of the need to take seriously into account the constitutive and irreducible differences within cultures—or, in other words, within systems—that provides a rather large stumbling block for this bioethics.

Although seemingly developed with the best of intentions at heart, since it is presented as driven by his desire for "the survival and improvement of mankind,"[55] Potter's ethics replicates the kinds of philosophical problems that Maturana and Varela originally ran against in *The Tree of Knowledge* in their own turn from cybernetics to ethics. It is precisely the maintenance of the system that Maturana and Varela positioned as an ethical imperative, just as Potter did, even though they acknowledged the radical contingency of observation and advocated vigilance against "the temptation of certainty." However, in their attempt to bring about the peaceful coexistence of the social system, they resorted to what Wolfe terms "unreconstructed humanism" by taking recourse to the idea of love as a guarantee of the functioning of their ethics. Believing that people would somehow naturally work out, in a rational manner, that acceptance of difference is better, they idealistically expected that "ethics may somehow do the work of politics"[56] by eliminating contingency and antagonism through careful and considerate reflection. Wolfe is also rather scathing about Maturana and Varela's turn to Buddhism (a turn that Potter also takes!), accusing them of trying to solve "by ethical fiat and spiritual bootstrapping the complex problems of social life conducted in conditions of material scarcity, economic inequality, and institutionalized discrimination of various forms."[57] Cybernetic "open systems" are suddenly starting to look rather closed, because they leave historical contingency as well as the materiality of the actual political conditions on the outside—an accusation that could also be levied at Potter.[58] As Wolfe concludes, "Maturana and Varela's humanist ethics thus fails precisely because it *is* humanist. . . . [It] forgets what their epistemology knows: that in the cyborg cultural context of OncoMouse and hybrids of nature/culture, the question is not who will get to be human, but what kinds of couplings across the humanist divide are possible and indeed unavoidable when we begin to observe the end of Man."[59]

Deciding on the System (from Outside the System)

In all its potential loosening up of human agency and the exploration of the constitutive relationship between the human and his environment, Maturana, Varela, and Potter only manage to outline a technical program for the betterment of the world, not an ethics. The main two problems with their proposals lie for me, first, in the elimination of the moment of decision, with their bioethics only becoming a plan for the successful running of the predecided program and, second, in the foreclosure of its posthumanist promise, whereby the specificity of embodiment and its cultural and political positioning gives way to timeless structural analogies across systems and layers. This kind of ethics seems politically orthodox, even totalitarian, as it is primarily focused on maintaining the world as it is and ruling out a number of (nonsystemic) possibilities in advance. Even though Potter himself acknowledges the possibility of a mistake occurring within a system, he sees it as an integral part of the system itself, a reverberation that does not disrupt the system as a whole. It is precisely this inability or unwillingness not only to think the total breakdown of the system but also to explore in great detail what happens at its margins, what is excluded by it, and how its boundaries are consolidated and maintained that makes his bioethics ultimately nonethical. Indeed, Potter emphasizes that disorder has to be eliminated to improve the human condition while knowing in advance (through consultation with the aforementioned "experts") what it is that humans need for their survival and well-being. Thus, even though his proposal seems to have sprung from genuine care about the fate of humans, their environment, and the world at large, Potter actually withdraws from proposing an ethics and gives us instead a program for repairing the world.

Potter's bioethics ends up being non- or counterethical because it forecloses on the need for a decision that would be heterogeneous to knowledge and that would exceed the know-how of the system. It thus forecloses on undecidability—which, arguably, is a key condition of ethics. In an oft-cited passage, Jacques Derrida claims that "If you don't experience some undecidability, then the decision would simply be the application of a programme, the consequence of a premiss or of a matrix."[60] Interestingly, Derrida resorts to computing terms in order to explain how ethics operates. He does acknowledge that an element of calculation is necessary in politics and that ethics is always already tied to the need for a political response to conflicting demands. We could perhaps even say that ethical decision involves having to choose between zeros and ones. However, Derrida makes it also very clear that, in order to actively make a decision between "two determined solutions" which are "as justifiable as one another"—rather than allowing for a program to execute itself—we need to go beyond the known sequence of zeros and ones. A decision thus involves going beyond the system; it is a moment of madness in the face of knowledge which is external to that knowledge; it must be an act of faith, an eruption of a ghost in the machine. And thus if we are to think ethically, we must consider the possibility that the system may collapse, that the world as we know it will explode, that, say, communism will offer a viable political or intellectual alternative (something that Potter rules out from the start),[61] that another set of values which will not be focused on individualism, human dignity, and prosperity will emerge. This is not to say that such outcomes are to be seen a priori as more interesting or better alternatives to the values that Potter builds his bioethics on; it is only to suggest that they need to be at least considered, not excluded in advance. Cybernetic ethics as devised by Potter insists on closing the gap between the "is" and the "ought," but ethics, we may argue, if it is to be fully ethical rather than just "mathematical," has to decide on the "ought" every time anew, without the guarantee of the fixed "is," or the possibility of the ultimate closure.

Breaching Totality (with Levinas): A Nonsystemic Bioethics of Relations

I thus want to suggest that if bioethics is to be conceptualized as ethics at all, it needs to ask very different questions from those posed by cybernetics-driven ethics: What happens when the system fails? Is there anything outside of the system? Is there anything that is excluded from it? How are the system's boundaries maintained? An ethical bioethics (a pleonasm to which I hesitantly resort in order to differentiate my "alternative bioethics" project from Potter's) needs to respond to the moment when the system is getting out of sync, when it is unworking itself; it needs to be attuned to the singularities within the system that cannot be reduced to mere genetic and cultural variations of a species. I should perhaps also explain that by insisting on an outer-systemic moment of madness and leap of faith, I am not bringing back humanism to cybernethics—even if I do recognize cultural, political, and ethical differences not only between humans, animals, and machines (temporary and unstable as these categories may be) but also among *different* humans, *different* animals, and *different* machines. The nature and significance of the ethical difference between, for example, "abusing a dog and abusing a scallop"[62] will nevertheless have to be responsibly decided always anew, in particular contexts, networks, and environments we will find ourselves in, and with all the knowledge and affect we will be able to mobilize.

All these reservations notwithstanding, Potter does offer us a possibility of envisaging a different bioethical model, one which looks at the biosphere as a non–hierarchically differentiated whole and that foregrounds relationality between different species and material forms.[63] Cybernetics can thus help us in our search for "doing bioethics otherwise," even if this "new" bioethics I am attempting to outline will have a very different relation to "the system" as such. This is by no means to position it outside of the system but rather to assign it the role of a system's bug: always attentive, tracing the networks, relations, and connections, and detecting unexpected openings and cracks within the system. Derrida instructs us that this very possibility of finding a metaphysical trace "on the other side" of the system is already inherent in cybernetics, because if it is "by itself to oust all metaphysical concepts—including the concepts of soul, of life, of value, of choice, of memory—which until recently used to separate the machine from man, [the theory of cybernetics] must conserve the notion of writing, trace, grammè [written mark], or grapheme, until its own historico-metaphysical character is exposed."[64] Cybernetics therefore cannot escape its own entanglement in the system of language, of writing, and hence in difference (of identity, signification, and meaning), even if its proponents position information as content- and meaning-free.

The cybernetic proposition that "there is no there there" and that the observer cannot stand outside the system being observed foregrounds the constitutional role of difference (i.e., all this that is not "the subject") in bringing forth this subject and establishing a relationship between the subject and its difference. We can perhaps trace a similarity here between a cybernetic understanding of subject formation and Emmanuel Levinas's notion of ethics as recognition of, and response to, the infinite alterity of the other. Levinas's philosophy emerges from Husserl's phenomenology, which connects the production of meaning with our lived experience, while postulating that our consciousness is an intentionality which is always in contact with objects outside of itself. For Levinas, phenomenology is "a way of becoming aware of where we are in the world,"[65] or, to put it differently, of what position we occupy within the system. However, it is not only discovering our place in the world but also accounting for it that constitutes the main premise of Levinas's thought.

My “place in the sun” is for Levinas always a usurpation; it is never *originally* mine. Instead, it belongs to the other whom I may have oppressed, starved, or driven away. No matter how much we invest in the illusion of our own self-sufficiency and power, for Levinas we always find ourselves standing before the face of the other, which is both our accusation and a source of our ethical responsibility.

It may be tempting to just reduce Levinas’s thought to a politics of guilt, echoing the Judeo–Christian obsession with disobedience, sin, and God’s wrath. And yet, even though the Judaic tradition does inform Levinas’s work in more than one way, and even though references to the idea of God feature prominently in his writings, God is not the *principle* of ethics for Levinas, nor does He provide a positive content in Levinas’s ethical theory. Rather than as a religious project, Levinas’s work should be seen as, first of all, an opening within the edifice of Western philosophy, or as a way of thinking that poses a challenge to the traditional (Greek) mode of philosophizing which shapes our concepts of time, space, truth, good, and evil. What Levinas is attempting to achieve in his writings—predominantly in his two major volumes *Totality and Infinity* and *Otherwise than Being*—is a postulation of ethics as a first philosophy, situated *before* ontology. This “before” should not be understood in a temporal sense: from a linear temporal perspective ontology cannot be preceded, as there *is* nothing before being. However, ontology as a “philosophy of being” is seen by Levinas as a “philosophy of power” and “injustice”[66] that tries to reduce any idea of the other to the terms and categories possessed by the same (which amounts to describing to what extent the other *is* or *is not* like me). Ethics thus becomes for him a different mode of thinking, one which “precedes” ontology in its relation to knowledge and justice. Instead of attempting to thematize and conceptualize the other as always already known, ethics points to the radical and absolute alterity of the other which collapses the familiar order of Being and calls the self to respond to this alterity. This possibility, as well as necessity, of responding to what Levinas defines as an incalculable alterity of the other is the source of an ethical sentiment. Refusing to assert the self’s primacy, Levinas focuses on the vulnerability of the self when facing the other, who is always already “absolutely other,” and who cannot be fully grasped by the self. This otherness can evoke different reactions in the self. But even though the other can be ignored, scorned, or even annihilated, he or she has to be first of all addressed, that is, responded to, in one way or another. This turning toward and response to the other, even if performed in the greatest arrogance and self-illusion about one's originality and self-sufficiency, is also already a welcome. It is an opening to, and a recognition of, the other, of his destitution, wretchedness, and suffering, but also of his infinite alterity. It is in this sense that ethics for Levinas is primordial, and that it is inevitable.

Let me reiterate here that this inevitability does not mean that the other cannot be annihilated. It is precisely the possibility of killing the other, and the fact that "the Other is the sole being I can wish to kill"[67] that make the relationship between us ethical. The *physical* possibility of taking the other's life does not therefore contradict what Levinas describes as "an *ethical* impossibility of murder."[68] If "Murder exercises power over what escapes power" and if "I can wish to kill only an existent absolutely independent, which exceeds my powers infinitely," my attempt at the total negation of Infinity only confirms it.[69] Levinas's philosophy is thus a blow to human self-aggrandizement; it is a philosophy of humility, but not in a traditional Christian sense, more in a skeptical sense of the recognition of one's limitations, as well as one's embeddedness within the system of the world. We may, of course, ignore these limitations, but to do so would be both tragic and foolish. It is the relation with the other that "introduces me to what was not in me" and it is the other that founds my idea of reason. In this way, the other allows for a passage from the "brutal spontaneity of one's immanent destiny" to the idea of infinity which "conditions nonviolence" and thus "establishes ethics."[70]

Levinas's ethical thought is helpful in providing a framework and a justification for caring about the life, any life, of the other, especially the precarious and destitute lives of all those who lack recognition in the dominant political discourses and policies, and those whose biological and political existence is confined to "zones of exception": comatose patients, asylum seekers, refugees, people with nonnormative bodies and looks, victims of biotechnological experimentation. It also transforms relationality between sentient beings from an objective material fact perceived as such from any point within the system to a subjective infinite responsibility directed, nonsymetrically, only at me. However, drawing on Levinas in an effort to develop a nonhumanist bioethics is not unproblematic. As John Llewelyn observes, "Levinas seems to imply that I can have responsibilities only toward beings capable of having responsibilities,"[71] that is, beings with whom I can be face to face, with whom I can enter into a discourse. Hence, it is only really human others with whom I can maintain an ethical relationship—and this, in spite of a certain opening made by Levinas himself in the essay "The Name of a Dog, or Natural Rights," where he writes about the dog Bobby, greeting Jewish prisoners in the German POW camp returning after a day's labor, *as if they were human.* Even though Bobby, "the last Kantian in the Nazi Germany," lacks "the brain needed to universalize maxims and drives," at the hour of trial it can "attest to the dignity of its person."[72] Bobby therefore signifies, in Levinas's own words, and with more generosity than Llewelyn perhaps allows for, "the debt that is always open."[73]

However, even if Levinas does hint at the possibility of a debt to the animal other, his ethical theory undoubtedly suffers from an anthropological bias, which is evident in the excessive weighting he gives to the logos, or rather perhaps in the too narrow conception of the logos as human word, as discourse between equals. I want to suggest, however, that Levinas's "error" is first of all scientific and historical rather than philosophical per se, in that he does not consider seriously the limitations of his own concept of the human as a speaking being with the face, rather than a sentient being reaching to—and touched by—the other in a myriad different ways. His commitment to the difference and singularity of the human seems to blind him to the discursive limitations of his own language through which ethical responsibility toward the other is justified. Because do we really know with whom we can enter into a discourse (a refugee? a cat? a computer bot?) and what this "entering into a discourse" actually means? And yet, the recognition of these limitations of Levinas's philosophy does not take away from the significance of his understanding of ethics as responsibility for the alterity of the other, nor does it invalidate his work for our project of thinking bioethics otherwise, even if it does make the task of encountering and dealing with this (human and nonhuman) alterity much more complicated. Cybernetics and systems theory can help here in offering a supplement to the humanist limitations of Levinas's thought in allowing us to envisage and take stock of humans' relationality and kinship with nonhuman others. However, it is through Levinas that we can take a peek at "the other side" of the system, which I understand less as transcendence and more as a pragmatic possibility of tracing the system's limits, of breaking through the system's totality.

Levinas himself draws on the language of systems to explain the emergence of signification in the self–other relationship. He recognizes the need for an effort which has to be made "for the structures to be packed in" to the system, an act that is to ensure intelligibility and reveal being in its essence. The "tight packing in of structures" is a condition of signification as well as a guarantee of the maintenance of a totality of the system (the point Derrida takes up in his work, arguing that his dictum, "there is nothing outside of the text," can actually mean that there *is* (almost) nothing outside of the system of signification, in the sense that we cannot identify any present or perceptible limit to this system).[74] Levinas writes: "It is because the assembling of nonsignifying elements into a structure and the arrangement of structures into systems or into a totality involves chances or delays, and something like good or bad luck, because the finitude of being is not only due to the fate that destines the way it carries onto manifestation, but also to the vicissitudes and risks of a packing in of its manifested aspects, that subjectivity in retention, memory and history, intervenes to hasten the assembling, to confer more chances for the packing in, to unite the elements into a present, to re-present them."[75] The subject is thus part of the "system" of being; it is "part of the way being carries on,"[76] but its presence and re-presentation are only temporary. They are also predicated on chance, on good or bad luck. The system's totality is for Levinas only ever provisional. Indeed, he is more interested in what we could describe as "an accident in the system," an unexpected state of events that becomes an ethical opening.

Any attempt to theorize a nontotalizing cybernetics-inspired bioethics needs to take into account the limitations of, and openings to, its own theoretical systematicity. What we need is a form of theorizing which would go against the visual mastery of seeing as much as one can, a mastery which is implied in the Greek roots of the term "theory" (*theorein*—to look). An ethical theory which would be truly ethical, and which would adopt a responsible stand toward difference, would need to break through any systematic totality, starting instead, to quote Levinas, from "a 'vision' without image," "a signification without a context."[77] He writes in *Totality and Infinity*: "The absolutely other, whose alterity is overcome in the philosophy of immanence on the allegedly common plane of history, maintains his transcendence in the midst of history. The same is essentially identification within the diverse, or history, or system. It is not I who resist the system . . .; it is the other.[78] For Levinas, ethics is not just an operation of thought—but neither is it arational or thought-free. It happens at the limits of thought. We could describe it as a leap of faith beyond knowledge, but undertaken *from* the position of knowledge. The operation of knowledge and critical thinking would not need to cease, though; instead, it would involve thinking that *there is unthinkable*, "an other refractory to categories," even if we can never ever ultimately think, that is, grasp, this other. (To put it in terms elaborated by systems theory, it would involve being able to think the inability to see the social or natural system as a totality from any particular observer's point of view.) Arguably, ethics can thus be positioned as taking place in a crack within, or even in the moment of resistance against, the system. It involves a breach of the systemic totality by infinite alterity, but not a passage "to the other side," as that would only amount to absolute immanence being replaced with absolute transcendence and would actually confirm totalizing identitarian categories.

The other, as explained by Derrida in his readings of Levinas in "Violence and Metaphysics," can never be "totally other." There is always some sort of relationality to the "infinitely other," a trace of difference in the same. It is precisely in this tension between categories such as self–other, inside–outside, and immanence–transcendence that a diachronic ethical relation with alterity is established. And it is in discourse—what Levinas calls speaking and what cybernetics refers to more broadly as communication—that ethics takes place. Of course, not all forms of communication are ethical per se. Indeed, in Wiener's conceptualization of cybernetics, communication is always coupled with control, and thus, if interpreted from the Levinasian perspective, it would be seen as *un*ethical. The paradox of Levinas's ethical proposition lies in noticing this possibility of ethics included in any form of communication, which is always a turning toward the other, an (as yet content-free) address, a welcome and a reaching out. Because it is a paradox, this welcome can turn out to be not very welcoming at all, but hostility will occur only *after* the welcome, that is, the reaching or turning toward, has taken place. What is needed for communication to be potentially ethical is this very *possibility of an address*, of turning toward. It is the eventfulness of the event—the fact that *something takes place between me and the never fully knowable other*, that a relation is established and entered into—that is significant here. The Australian artist Stelarc playfully illustrates this point in a conversation with his Prosthetic Head, a performance agent equipped with a text-to-speech engine with a source code for facial expression and real-time lip-syncing:

STELARC: "Who are you?"

PROSTHETIC HEAD: "That's not a meaningful question. What is important is what happens between you and me. It's what happens in the space between us that matters. In the medium of language within which we communicate, in the culture within which we've been conditioned at this point in time in our history."[79]

Some readers may be concerned at this point whether this does not actually reinstate "old" humanism in this "new" ethics, where machines and animals are still "like humans" and where it is the human who ultimately makes all the cognitive and ethical decisions. Yet I suggest that a Levinas-inspired ethics of relationality, of turning toward, goes beyond the traditional humanist tenets because something very radical happens here to the very idea of the human when it is analyzed in terms posited by cybernetics (i.e., as being part of the system and thus dependent on the system's technologies). The one-for-the-other structure of subjectivity in Levinas, which is for him "a conjuncture in which a man is responsible for other men,"[80] needs to be expanded if under the conditions of digitality we cannot ascertain any longer whether the other who is "assigned to me" and for whom I am responsible is human or machinic, and whether the "fraternity" Levinas talks about extends to all of DNA kin (chimpanzees, dogs, bacteria). As Stiegler puts it,

> I am already and have always been constituted by my relation to the *mēkhanē* and, through it, to all possible machines. It was around four million years ago, according to the dating of Leroi-Gourhan (since reduced to 2.8 million years ago), that a new form of life appeared, one supported by prostheses. . . . This living being that we call man and that this myth [of Prometheus and Epimetheus] soberly designates as *mortal* (that is to say: the being who anticipates his own end and his *difference* from the immortals, from whom he receives, albeit by theft, his power, his fire, that is to say *tekhnē* and all its possibilities, and who therefore endures, in the ordeal of non-predestination that results from this, the experience of a difference marking his origin, which is thus essentially the difference between the sacred and the profane) is a being that, to survive, requires non-living organs.[81]

The human is thus for Stiegler always already technological: he is coconstituted with and through technology (a flint, say) and depends on *tekhnē*, *mēkhanē*, and "non-living organs" for his survival. At the same time, it could perhaps be argued that the digital age exacerbates this technological condition of the human to an unprecedented extent, with technologies and media becoming part of the human in a much more ubiquitous, everyday, and embodied way. With mobile phones, iPods, wireless Internet connections, immersive game environments, and the convergence of different media forms and contents, the lines between what used to be more comfortably recognized as "real" and "virtual" relations are becoming increasingly blurred.[82] The human is positioned much more ostensibly as an element in the information system, a nodal point for the flow of data, rather than a skin-bound, self-contained rational moral agent. Even if we were to argue that this is a change of degree rather than of kind, "the degree of this degree" in the digital era is significant enough to call, together with Mark Poster, for "an innovation in the theory of ethics" and thus for the development of "new ethical rules . . . for mediated culture."[83] Indeed, for "humachines," the very nature of what I called "relationality" becomes much more unstable and fluid and much less dyadic than it is in more human-centered ethical positions.

Although it can learn its posthumanist lesson from cybernetics, any nonhumanist ethics worth its (ethical) salt will not entirely remove qualitative differences between humans, animals, and machines, nor will it replace the multiple differences between species and life forms with a species continuum, or a seamless life (or data) flow.

So what is the nature of these differences? I do not know.

Now, I would like the reader to consider with me the possibility that what might perhaps seem like a flippant answer, an "intellectual cop-out" which only ends up reaffirming humanism via the human's ability to doubt, is actually a serious philosophical proposition for articulating the relationship between humans, animals, and machines. Indeed, anything else—no matter if I were to defend the special positioning of the human as a being with its own teleology and truth, or the species continuum of modern naturalism which only affirms differences of degree, not of kind—would require the reinstatement of the position of knowing the nature of this difference and being able to arbitrate over it once and for all. Simon Glendinning argues that a Derrida-inspired position of *dwelling on this difference* will allow us to avoid the "fundamental *cognitivism* with regard to proper appreciation of the human difference" that classical humanism and modern naturalism share. For Glendinning, both of these theories of the human claim that a proper grasp of the significance of human difference "is ultimately, *decisively*, a matter of our having adjusted our beliefs to how things really are."[84] The *ethical* recognition of this difference, in turn, does not amount to knowing its nature once and for all. Indeed, any attempt to cognitively master this difference will have, as Stiegler postulates, a mythical character—less in the sense that it is opposed to a true story waiting to be discovered and more as a narrative that is yet another one of the prostheses that constitute our humachinic existence. In other words, myths make us (human) as much as flint tools do.

Also, the question whether this ethics also applies to others—other humans but perhaps also apes, dolphins, or even "intelligent machines"—is not really important, because *it only ever applies to me*. It is *my* anxiety about death and *my* awareness of my own mortality that establish a temporality for me while also opening up a set of possibilities.[85] These possibilities always include that ultimate possibility of an end—the end of me as a specimen of "humans" and the end of the human as a species.[86] As soon as I start asking questions about the general applicability and universal validity of this ethics, I am introducing calculation into it by trying to measure my responsibility against that of (human and nonhuman) others. Of course, it is understandable that I may *want* to do this, but in this way I am already moving from ethics into politics, the domain of making strategic decisions and arbitrating between different procedures and events. Again, these decisions have to be made, every day, in the material conditions in which we find ourselves and against the particular social, economic, and cultural circumstances we are embedded in. However, the political questions we will ask about the relations between humans, animals, and machines will be very different from the ethical injunction that the human or nonhuman other—who is both my kin and infinitely different from me—makes upon me and that precedes any political action. It is precisely in the resistance against the fusion of this relationality between me and the other that ethics is enacted. Because "It is not I who resist the system . . .; it is the other."

3

Biopolitics Today: The Ubiquitous Practice of Life Management

June 08, 2007
It's no secret that I've been having a hard time of it lately. You don't even need to click back through the archives, you can just scroll down the screen to witness that I'm out looking out at the abyss. For the past week I've been trying to figure out where I screwed up, how I can pinpoint the exact moment I made the wrong decision. I'm a big believer in taking responsibility for your life and I've been trying to take responsibility for mine—and now Will's—to figure out how to get out of the financial and career mess that I've gotten myself into. *I have to contribute* is the mantra that's on eternal repeat in my head, playing faster and faster and louder and louder until the pressure starts building and I can feel the pinpricks of the migraine beginning to jab at the back of my eyeballs. I scan job boards and send out resumes and make follow up phone calls and try to get a temp agency to even give me the time of day, as Los Angeles has a glut of unemployed overeducated workers all scrambling for the ten-buck-an-hour temp job.
The Slack Daily
http://www.theslackdaily.com/

The Politics of Life Management

As the reader should be aware by now, the conventional definition of bioethics as a discipline dealing with "ethical issues arising from the biological and medical sciences"[1] is broadened in this volume to refer instead to something we might describe as "an ethics of life." However, since "life" is a rather nebulous concept with extensive metaphysical connotations, and since it functions as an object of study within disciplines as diverse as philosophy, literature, anthropology, biology, and computer science, a reservation perhaps needs to be made that this "ethics of life" does not stand for an "ethics of just anything." Rather, bioethics, as it is positioned in this book, engages with life in its Greek etymology conveyed by the terms *zoē* and *bios*, that is, with life in its biological and political aspects. In other words, the book embraces life as a network of material and symbolic forces that are in operation in the world and that shape both our metaphysical and technological concepts and paradigms. Conventional bioethics has typically been more preoccupied with the *zoē* aspect of life, that is, the "raw," biological life of individual organisms (hospital patients, lab animals, genetically modified species), rather than with their political existence (*bios*). Yet when we are faced with singular decisions concerning individual beings, and their lives and health issues, we are always already situated in, and drawing on, a broader political context. A bioethical decision is therefore not just moral but also political in its not always acknowledged motivations. It also has political consequences.

If bioethics, even in its "expanded" definition discussed above, is principally concerned with the care of life, it could be argued that the very process of defining and managing this life—on a micro- and macro-level, in scientific laboratories and sociopolitical institutions—constitutes the predominant political question today. One of the key tenets of this chapter is that politics in Western democracies has now to a large extent taken the form of biopolitics, a political regime under which bodies and minds of citizens are administered and under which life is "managed." Drawing on the work of Michel Foucault and Giorgio Agamben, I will argue that this process of biopolitical "life management" involves the constitution of zones of indistinction between political and biological life, which also entails barring some people and other forms of life from the democratic polis (e.g., asylum seekers polluting the healthy "body politic," prisoners kept in the Guantánamo Bay camp outside national or international jurisdiction, genetic in-valids as prefigured by Andrew Niccol in his 1997 movie *Gattaca*). The indistinction between political and biological life has serious ethical implications. My thesis is this chapter is thus twofold:

1. Bioethics operates in a biopolitical context.
2. Even though biopolitics as a process of "life management" must not be designated as "wrong" or "bad" in advance, and should rather be

understood as a set of relations and forces that enable both domination and freedom, it needs a clearly articulated ethical supplement. The role of this bioethical supplement is to counteract the anti-ethical moralism[2] and the profit-driven economism that drive many political decisions today and that legislate between legitimate and illegitimate life forms and life modes.

In contradistinction to the traditional bioethical positions outlined in chapter 1, which are grounded in analytical philosophy and clinical practice, the alternative bioethical framework I aim to develop throughout the book is rooted in the Levinas-inspired understanding of ethics as existing prior to politics and as making a demand on politics (although this framework will have to go beyond the humanist limitations of Levinas's own philosophy). The priority of ethics springs from the fact that the question of the management and regulation of life always already occurs against the horizon of obligation exuded by different life forms. This obligation consists in a need to respond *in some way* to a phenomenological and affective impression that the "touch of life" exercises upon bodies and minds. The obligation is in itself content-free, in the sense that "life" is not posited in advance as the highest value that should be protected at all cost. The nature and direction of this ethical response is therefore not determined a priori, although in particular circumstances and from particular ideological positions this content-free obligation does get filled with specific values. (These values may lead one to assert, for example, that the precarious life of humans in war-torn Iraq needs protection or that industrialization and the enhancement of the quality of "human life" does not justify wiping out more than a quarter of animal populations on the planet.) Driven by the need to provide a response to the alterity of other beings and life forms, and for a decision, always to be made anew, about what to do, a bioethics informed by the philosophy of difference retains a sense of responsibility, even if the locus of this responsibility, response, and decision is not unequivocally located in an individual, rational self.

The Levinasian conceptualization of ethics as hospitality and welcome, where the ethical drive comes from outside the self, will be confronted in this chapter with the Foucauldian understanding of ethics as aesthetics, a responsible project of self-creation. Through an engagement with specific practices of life writing and self-creation in the age of new media, such as blogging and social networking (via portals such as Blogger, LiveJournal, or MySpace), I will consider to what extent an encounter between Foucault's ethics as "the care of the self" and Levinas's ethics as "responsibility for the alterity of the other" is possible or even necessary in the current cultural conjuncture. I will argue that Levinas and Foucault allow us to understand what the author of The Slack Daily blog quoted in the epigraph to this chapter calls "taking responsibility for your life" as something more than just an individualistic neoliberal imperative to align one's aims with those of the current politico-affective economy. In my attempt to outline a different bioethical framework via the analysis of blogging, I also hope in this chapter to go some way toward blurring the distinction between ethical theory and ethical practice. The chapter is thus intended as a bridge to the "Bioethics in Action" part of the book that follows.

First of all, a few more words about biopolitics, a concept that is being increasingly used by cultural theorists to describe the working of political regimes in the West. Its theoretical relevance notwithstanding, the widespread use of this concept is no doubt related to the 2003 English translation of Foucault's *"Society Must Be Defended,"* in which power and politics are analyzed through a "bio-framework." The book, which follows Foucault's earlier studies of power in disciplinary institutions such as the prison, the army barracks, the hospital, and the asylum, contains a series of lectures delivered by him between January and March 1976. It examines a shift in the working of power, from disciplinary power exercised on individual bodies to biopower exercised on whole populations. Foucault also takes up this concept of biopower, which he links with the idea of biopolitics, in the first volume of *The History of Sexuality*, published in December 1976. He traces back the origins of this regime of biopower to the classical rule of power over life and death, whereby the sovereign exercised "his right of life only by exercising his right to kill, or by refraining from killing."[3] The sovereign's "power of life and death" amounted to the right to take the life of his subjects and to allow them to live. Power at the time operated by deduction and subtraction—of life and its material realizations, such as things and bodies. In the nineteenth century, however, sovereign power underwent a process of transformation, with emphasis placed on the positive management of the life of all citizens rather than just on the defense of the sovereign. The sovereign's authority has since then been focused on "making" live and "letting" die,[4] that is, on "augmenting" rather than "subtracting" life. Although sustaining and enhancing life has become the highest prerogative of modern power, war, genocide, and putting to death are still seen as a necessary, if undesirable, part of this regime of "life management."

Foucault distinguishes two levels on which this regime of life and death management has been operating since the seventeenth century. The first one of these is the level of individual bodies, where the body is conceptualized as a machine: "its disciplining, the optimization of its capabilities, the extortion of its forces, the parallel increase of its usefulness and its docility, its integration into systems of efficient and economic controls, all this [is] ensured by the procedures of power that characterize[. . .] the *disciplines*: an *anatomo-politics of the human body*."[5] The second level of power is directed at "a multiplicity of men . . . to the extent that they form . . . a global mass that is affected by overall processes characteristic of birth, death, production, illness, and so on."[6] In other words, it works on the species body, that is, the population, and takes care of issues such as propagation, births and mortality, life expectancy, and general health, with the care aspect of life management overshadowing, if not entirely replacing, its disciplinary aspect. Medicine becomes one of the techniques through which power is exercised not just over individual bodies but also over bodies en masse, with increased focus on public hygiene, accidents, infirmities, and various anomalies, as well as issues connected with reproduction (abortion, genetic counseling, sterilization, etc.). This current medicalization of populations is accompanied by their "digitization," in the sense that various numerical, statistical, and informatic means of managing and "regularizing" groups of bodies are being established and widely implemented. Foucault terms the regulation of these issues at a population level "*a biopolitics of the population*," although we could argue that these two levels of life management are interpenetrated and that biopolitics entails an anatomo-politics, that is, the disciplining of the body–machine, the constant manipulation of the mechanics of life.

Biopolitics has far-reaching ambitions: it adopts "control over relations between the human race, or human beings insofar as they are a species, insofar as they are living beings, and their environment, the milieu in which they live."[7] We should bear in mind Giorgio Agamben's argument that the Greek etymology of the word *bios*, which in common parlance stands for the material "stuff" of life, actually points to "a form or way of living proper to an individual or a group."[8] Wherever there is life that is more commonly identified as biological, the sheer fact of living, or what Agamben designates as *zoē*, "there is already population and organizing potential."[9] Life is always already multiple, "bare life" common to all living beings is, at least potentially, *organized*, that is, *political*, existence. Indeed, according to Agamben, biological life occupies "the very center of the political scene of modernity."[10] Also, politics is *constitutive of* the human as a living being rather than being merely an *addition to* human raw matter. From this vantage point, the population can be described as both a political and a material–biological entity, consisting of machinic elements which are endowed with forces of life and which need to be properly administered. It can be argued that modern politics takes place in the tension between the discourses and practices of disciplinarity, which become less pronounced in contemporary democracies, and those of care, which are now much more explicit. It is also from within this tension that a possibility of resistance against the constraining disciplinary forces can arise. This possibility, as Foucault indicates, emerges at the level of actual living bodies, whose docility and disciplinary submission are never final: bodies inhere potential for rearranging the vital forces that situate them in the actual networks of power and, thus, for shifting the existent architecture of power.

Life in the State of Exception

Before we move on to resistance and the possibility of ethical transformation from within the existent power structures, let us explore a little further the more constraining aspects of biopolitics. Drawing on

Foucault, in *Homo Sacer* Agamben argues that violence is inherent to sovereign power. He also identifies the concentration camp and the great totalitarian state of the twentieth century as two exemplary places of modern biopolitics.[11] In either of these places, what Agamben calls "bare life" (*la nuda vita*, the sheer fact of living prior to articulation and organization that he also describes with the Greek term *zoē*) enters the political realm in most explicit ways, but this entry is not straightforward—it only takes place as an exception. In Auschwitz, for example, "the Jew" as a biological entity also fulfils the *negative* political function of demarcating the boundaries of legitimate citizenship, while the *Muselmann*, "the living dead," functions as a *negative* threshold between life and/as labor in the camp, on the one hand, and biological and political finality on the other. In this sense, bare life "presents itself as what is included by means of an exclusion."[12]

With this argument Agamben challenges Foucault's claim that modern biopolitics is a new form of politics that constitutes a decisive break from classical political formations. He also disagrees with Foucault's conclusion that sovereignty is losing its political purchase in the era of biopolitics. Agamben in fact radicalizes Foucault's concept of biopolitics by arguing that it is not really the inclusion of *zoē* (raw, bare life, or the simple fact of living) in the polis or even the fact that "life itself" becomes a principal object of the calculations of state power[13] that is most significant about modern politics. What is even more important for Agamben is that in modern democracies "the realm of bare life—which is originally situated at the margins of the political order—gradually begins to coincide with the political realm, and exclusion and inclusion, outside and inside, *bios* and *zoē*, right and fact, enter into a zone of irreducible indistinction."[14] In contemporary politics the original exclusion of *zoē* from the polis (which, as shown above, is actually "an inclusive exclusion," an *exceptio*) becomes extended and generalized to the point where it becomes the rule, a state of events which Agamben describes as "the fundamental political structure" today.[15] In an article written for *Le Monde* in 2004, he quotes "bio-political tattooing," that is, being fingerprinted and iris-scanned and then having one's biometric data deposited in a digital database when entering the United States, as one example of this exceptional procedure. Fingerprinting used to be "imposed on criminals and political defendants"[16] but is now being extended to citizens of all states who physically and politically find themselves on the threshold of the United States. In the light of the "global war on terror" post-9/11, the establishment of the "terrorist" detainment camps such as those in Guantánamo Bay, which are exempt from international jurisdiction, and the development of stringent immigration policies which involve the setting up of asylum seeker detention centers both within and outside the borders of the European Union, the list of the exemplary places of modern biopolitics proposed by Agamben in his 1995 book *Homo Sacer* needs to be revised and expanded. The inherently violent structure of the political, which Agamben identifies in the concentration camp and the totalitarian state, seems to have been transferred to both the globalized and localized spaces of governance in the twenty-first century.

In *State of Exception* Agamben himself extends the biopolitical paradigm to the "military order" instituted by the president of the United States in the aftermath of 9/11—a paradoxical state of events in which law is suspended by the force of law, leaving in operation "a force of law without law."[17] He explains this new political conjuncture as follows:

> What is new about President Bush's order is that it radically erases any legal status of the individual, thus producing a legally unnamable and unclassifiable being. Not only do the Taliban captured in Afghanistan not enjoy the status of POWs as defined by the Geneva convention, they do not even have the status of persons charged with a crime according to American laws. Neither prisoners nor persons accused, but simply "detainees," they are the object of a pure de facto rule, of a detention that is indefinite not only in the temporal sense but in its very nature as well, since it is entirely removed from the law and from judicial oversight. The only thing to which it could possibly be compared is the legal situation of the Jews in the Nazi *Lager* [camps], who, along with their citizenship, had lost every legal identity, but at least retained their identity as Jews. . . . [I]n the detainee at Guantánamo, bare life reaches its maximum indeterminacy.[18]

In this particular political setup the sovereign is perceived as someone "who decides on the state of exception," an idea Agamben borrows from the political theorist Carl Schmitt.[19] The sovereign has the authority not only to decide on the law but also to raise himself above the law and suspend it when what is seen as fundamental to the existence of all finds itself under threat. "This something is a notion of bare life, life that is, like the sovereign, at once inside the political order (in that the basics of life itself are protected within society) and also outside the political order (in that biological life serves as the natural basis for social and political life)."[20]

On the Generative Power of *Zoē*

Arguably, there is something rather disabling about this overarching biopolitical framework. Alison Ross, editor of the special issue of *The South Atlantic Quarterly* "The Agamben Effect," argues that Agamben's work operates in two different registers, which she respectively terms "diagnostic" and "promissory."[21] It is perhaps in the identification of the "high stakes the category of life carries in the West"[22] that Agamben's diagnosis is most powerful and most useful. The "promissory" aspect of Agamben's philosophy, in turn, seems to be left wanting to some extent. Many scholars have questioned where the transformative drive of any such future politics will actually come from and whether such a transformation is imaginable at all within the conceptual framework outlined by the Italian philosopher. One may also wonder whether the constant application of the tight interpretative paradigm provided by Foucault and developed by Agamben to numerous political events where lives are being institutionalized and managed, and where exception becomes the rule, even if intellectually seductive, does not in fact pose a danger of domesticating the new political situation we are faced with and thus actually atrophying political vigilance and critical spirit. By applying a biopolitical framework to the scene of international politics post-9/11, do cultural theorists not risk becoming complicit with the hegemonic, conservative articulation of what is going on in the world today, one that pronounces that "we" are now "at war" with the enemies of democracy and the West?

Some of these reservations are no doubt linked to the overt negativity of Agamben's notion of "bare life" within the biopolitical framework. It is debatable whether the exclusionary position that "bare life" occupies in Agamben's philosophy actually allows for any form of political agency to emerge and whether there is enough force, enough "life," left in it to warrant the reworking of negativity and the transformation of the political status quo. Rosi Braidotti, for example, argues that Agamben's fixation on the tragic aspects of modernity, with its totalitarian states and concentration camps, "fuels an affective economy of loss and melancholia at the heart of the subject" while also consolidating the imbalance of power between the all-potent sovereign and his weakened and passive victims.[23] In a similar vein, Ewa Ziarek criticizes Agamben for imposing conceptual and material constrains on the notion of bare life, which makes it impossible for it to be mobilized in the service of resistance and emancipation. She also points out the need to distinguish between "bare life—wounded, expendable, and endangered" and "biological *zoē*."[24] The former is only "the remainder of the destroyed political *bios*"; it is a disabled *zoē*. The conceptual slide between these two versions of life, and the absolute subjection of bare life to the impunity of the sovereign, seems to be an indication of the inability on Agamben's part to see vitality in embodiment: the latter presents itself to him as always already too violated and too immobilized.[25] This perhaps explains Agamben's failure to develop a more future-orientated life politics—even if his writings abound in pronouncements to the contrary.

And yet Ziarek insists on the very real possibility of recouping the vital political force of "bare life" in excluded bodies—for example, racialized bodies subject to slavery or gendered bodies experiencing political disenfranchisement through the denial of the right to vote. This vitality of enslaved bodies has to remain latent rather than be totally eliminated, she argues, because "the biopolitics of slavery is confronted with the profitable inclusion of socially dead beings."[26] This ambivalent situation of bare life in slavery, which hints at the slave owner's "parasitical dependence" on the life he denigrates and excludes, provides for Ziarek "a new ground on which to theorize the possibility of resistance and emancipation."[27] Using a similar tactic, she positions British suffragettes' actions in the hunger strike as a usurpation of the sovereign's power over bare life, and the resignification of that power as productive rather than merely oppressive. Ziarek's project is very much in alignment with other feminist propositions that take embodiment, vitality, and natality seriously and that, consequently, allow for a more optimistic reading of the biopolitical conjuncture. Braidotti seems to echo Ziarek's concern when she admits to departing "from the habit that favours the deployment of the problem of *bios/zoe* on the horizon of death, or at the liminal state of not-life," and stresses instead "the generative powers of *zoe*."[28] This turning toward the generativity of *zoē* will also allow us, in Braidotti's words, to counteract the "gloomy and pessimistic vision . . . of the technological developments that propel the regimes of bio-power."[29]

Indeed, it can be suggested that technological processes driven by digitization and informatization have given shape to a determinedly new mode of biopower and new form of biopolitics at the beginning of the twenty-first century. The conceptualization of "life" itself has undergone a significant transformation in what we may refer to as "the age of new media," where the processes of digitization are both ubiquitous and to a large extent invisible (i.e., we do not "see" what happens inside our computers or cell phones, we do not normally perceive the actual sequences of zeros and ones that are transmitted as data over the Internet, nor do we visually register the information-based processes occurring in our DNA). Through developments such as cloning, microelectronic implants in the retina to cure blindness, xenotransplantation, genetic interventions, artificial life, or the aforementioned biopolitical tattooing, digital technology and digital media have had a tremendous impact on the inherited concepts of the human and human life: they have challenged the idea of the human as a skin-bound singular entity while also opening up new forms of kinship between humans, animals, and machines. It is these particular developments that prompt a new articulation of what counts as biopolitics today, while also providing the ground for new bioethical thinking. The recognition of economic and ideological constraints at work in the technoscientific apparatus that shapes the current biopolitical regime does not have to blind us to the more productive aspects of technologies and media. Indeed, it is precisely from within the ambivalent dynamics of the technological relation—one which sets the human up as co-emerging with machines, tools, and media environments—that a different mobilization of life forces can be actuated and put into productive use, I want to suggest. The technologies that will interest me in particular in the further part of this chapter are those most explicitly connected with the production and articulation of "life" in the technological environment: that is, blogs and social networking portals, such as Blogger, MySpace, or LiveJournal. Via the study of these technologies, I intend to recoup the vitality of life itself and consider the productivity of technologies and media in which our lives and bodies are implicated. This will allow me to focus on a more "promissory" aspect of the biopolitical conjuncture and to work some way toward sketching out a bioethical supplement to the current biopolitical normativity. In order to achieve this, I suggest we temporarily depart from Agamben's theorization of biopolitics and "bare life" and return instead to his predecessor, Michel Foucault, whose "ethics of the self" implies the vital generativity and self-creation that can provide an opening within, and a supplement to, biopolitical domination.

The Ethics of the Self

> Monday, June 04, 2007
> Like so many other liberal, East Coast, middle-class, childless, self-satisfied, vaguely self-aggrandizing, "Gore in '08" couples gone by, the hubby and I have recently purchased a Prius. She's a lovely vessel, strong and true, fit for humans and hounds alike. And she needs a name.
>
> *There's a Blog in My Throat*
> http://bloginthroat.blogspot.com/

The subject of ethics does not feature explicitly in any of Foucault's major works. It is only in short articles and interviews, gathered in the volume *Ethics: Subjectivity and Truth*, and then in the series of lectures given at the Collège de France in 1982 and published as *The Hermeneutics of the Subject*, that Foucault addressed the issue of power and resistance through the ethical processes of self-fashioning and self-creation. Foucault understands ethics in terms of ethos, which stands for practice, embodiment, and a style of life. It is a practice of freedom derived from the game of truth.[30] Its origins lie in the twin imperatives of ancient philosophy, "know thyself" and "care for thyself," which in Christian times undergo a transformation into the principle of asceticism. The care of the self has a critical function in this relationship between self-knowledge and attention to the self, as it enables one to "unlearn" bad habits and cast away false opinions. It can also play a curative role. Ethics thus becomes a vocation focused on realizing the twin classical imperatives in order to develop a free relationship of the self to itself, and on arriving at a "subject of desire" responsible for creating new forms of pleasure. The call for self-knowledge and the link between ethics and governmentality as "the way in which the behavior of a set of individuals became involved, more and more markedly, in the exercise of sovereign power"[31] position the ethics of the care of the self, especially in its Greek incarnation, as a communal task. Here, subject formation is to benefit others, but this value-free ethics is not a moralism that would require the performance of good deeds for the sake of other people. Instead, the care of the self focuses *on the self*, which is understood as a process and a project.

While in ancient Greece the care of the self was the consequence of a statutory situation of power, "entailed by and inferred from the individual's will to exercise political power over others,"[32] Foucault maintains that in Roman times, particularly in Epicurean and Stoic philosophy, it is repositioned as a permanent task to be taken up by every individual throughout his life—although there of course remain both political and individual limitations to this universalism. Significantly, ethics arising from the "care of the self" principle does not rest on any a priori rules of conduct but rather on the sense of obligation and duty that imposes itself, in a direct and often uncomfortable way, on us. Foucault explains: "The care of oneself is a sort of thorn which must be stuck in men's flesh, driven into their existence, and which is a principle of restlessness and movement, of continuous concern throughout life."[33] He also points out that in Hellenistic and Greco-Roman culture, *epimeleia heautou*—taking care of oneself—was a very powerful notion, which entailed constant working on oneself and was very different from "the Californian cult of the self," where you are supposed to discover "a true you."[34] The notion of truth is vital for Foucault, but he speaks about "particular truth" to be created by each self as a relationship to what is not in it, not a preexistent general truth already in place, waiting to be discovered. The self is connected to a truth outside of itself, which it has to work out, memorize, and progressively put into practice. For example, sexuality for Foucault is first of all something one creates, together with a whole "way of life" that accompanies it. In a similar way, "ethical living" hinted at, not without self-deprecating irony, by the author of

There's a Blog in My Throat quoted earlier on in this section, is established via a sequence of activities and practices, such as the purchase of an ecologically sound Toyota Prius. For these practices to "count" toward "ethical living," a truth narrative about them needs to be developed.

This idea of a particular truth to be arrived at and a particular way of life to be created foregrounds one of the most significant aspects of the ethics of the self: the absence from it of any prior codification or instructions on how to live. It is the ethical activity itself, "vigilant, continuous, applied [and] regular,"[35] that is important, rather than the fulfillment of any particular commandments. Foucault makes it clear that he is not outlining any kind of ethical prophetism with his ethics of the self when he insists that "People have to build their own ethics, taking as a point of departure the historical analysis, sociological analysis, and so on that one can provide for them."[36] The "transformation of one's self by one's own knowledge" and the elaboration of new, beautiful ways of life are of primary importance in this ethical project.[37] We can see here that Foucault's ethics has a strong aesthetic dimension. In his discussion of nonnormative sexualities he goes so far as to compare an imperative "to create a gay life. To become"[38] with the creation of works of art. Aesthetics does not, though, constitute for Foucault an autonomous sphere, separate from all other spheres of life. Instead, he revisits the ancient connection between beauty and moral worth, where art carries the more general sense of *tekhnē* (technique, skill) and *poièsis* (creation) and refers to more than just the production of beautiful objects.[39] According to Timothy O'Leary, ethical practice as a process of artistic creation for Foucault involves "giving a form to one's life through the use of certain techniques."[40] Ethics becomes equivalent here with "the art of life."

It is important to bear in mind that such self-creation does not occur in a vacuum: we cannot just become who we want to be as our very self-knowledge about our wishes and the direction of our transformation need to remain subject to a critical scrutiny. We are also bound by the power of institutions and social relations, and by their disciplinary and constraining effects. However, as Maurizio Lazzarato explains, for Foucault power is a strategic rather than a unilateral and oppressive relation, which means that every one, or, more precisely, every *body*, functions as a node in a network of forces that pass through it, with each force simultaneously constraining the body and endowing it with power. Here, power names "an integration, a coordination and determination of the relations between a multiplicity of forces."[41] It is precisely the resistance element of power relations, and the possibility of introducing change into them—via the rearrangement of the forces that pass through the body—that make Foucault's notion of power productive rather than just negative, and that create the possibility of agency for the subject on the way to its desire and truth. "The forces that resist and create are to be found in strategic relations and in the will of subjects who are virtually free to 'control the conduct of others,'" according to Lazzarato.[42]

In this project the self is established as a subject through "techniques of the self" (also translated into English as "technologies of the self"), "which permit individuals to effect by their own means, or with the help of others, a certain number of operations on their bodies and souls, thoughts, conduct, and way of being, so as to transform themselves in order to attain a certain state of happiness, purity, wisdom, perfection, or immortality."[43] Among these technologies, Foucault lists dream interpretation, "taking notes on oneself to be reread, writing treatises and letters to friends to help them, and keeping notebooks in order to reactivate for oneself the truths one needed."[44] It is his focus on letter writing and diary keeping that is particularly relevant for my attempt to assess to what extent social networking portals such as LiveJournal, MySpace, and Flickr as well as online diaries (blogs)—where writing techniques are being deployed for the construction of selfhood and agency, both on- and offline—could be read as facilitating the very care of the self Foucault analyzes in Greco–Roman culture and as perhaps even enacting a new ethical way of being in the world. However, before I proceed to address this issue, we need to ask whether we can actually transfer the ethics of the self developed in ancient times into the current historical conjuncture. Foucault himself is aware of the difficulty involved in this process of cross-cultural translation. And yet he finds enough material in the understanding and application of the care of the self in the texts of Plato, Plutarch, Seneca, Marcus Aurelius, and Epicurus to defend his genealogical travels between ancient Athens and Rome, and the late twentieth-century "West."[45] The pivotal point for this return to ancient philosophy is the similarity he identifies between the separation of ethics from the domains of both religion and law and the resulting impossibility of finding a positive principle on which ethics could be based in both ancient Greece and modern times. But rather than lament the decay of all values and the unfeasibility of ethics nowadays as a result of this "loss" of ethical foundation—as has been a tendency with certain critics of "postmodern nihilism"—Foucault upholds this content-free ethics as a key structure of existence, a lifelong responsibility and task.

Foucault is keen to emphasize the long-term scope of the care of the self project, which involves the formation of the self through techniques of living, as well as the permanence of the struggle it activates.[46] He argues that "the general Greek problem was not the *tekhnē* of the self, it was the *tekhnē* of life, the *tekhnē tou biou*, how to live."[47] Getting to the heart of his own biological and medical rhetoric, Foucault goes so far as to insist that "Permanent medical care is one of the central features of the care of the self. One must become the doctor of oneself."[48] I would therefore like to suggest that his ethics of the self can be mobilized for developing what I mean by *bio*ethics in this book, a performative ethics of life which both enacts new practices of living and looks after the life of its agents who are always on the way to their selfhood. I am aware that Foucault's work also yields itself to more traditional "bioethical applications,"[49] due to his extensive analyses of the power relations at work in hospitals, and can be used as a corrective to the excesses of disciplinarity exercised on the level of institutions and individual patients' bodies.

However, Foucault's work can allow us to do much more than merely repair the shortcomings of the existent bioethical discourses and practices concerning patient care and medical regulations. It is in itself already a form of bioethics, an alternative framework for thinking about life and ways of living it. Informed by the notions of the self and of life as always in the process of becoming, it can challenge the more stabilizing and constraining aspects of biopolitics discussed at the beginning of this chapter. It is important to emphasize that the ethics of the self must not be understood as an attempt on Foucault's part to promote a selfish, or worse, delusional, individualism under the guise of his *tekhnē tou biou*. The Stoic-inspired withdrawal from the affairs of the world should be seen instead as a better, more prudent preparation for being in the world and for relating to people, things, and events. As well as becoming an art of (whole) life, *epimeleia heautou* makes life better: it is said to have a curative and therapeutic function[50] and thus, as indicated earlier, remains in "kinship with medicine."[51] While in Plato the art of the body is clearly distinguished from the art of the soul, in the Epicureans and the Stoics "the body reemerges very clearly as an object of concern so that caring for the self involves taking care of both one's soul and one's body."[52] This "Foucauldian bioethics" presents a very different model of bioethics from the dominant one, which I outlined in chapter 1. In the latter, the self is most often positioned as a subject of medical procedures and political interventions. Even in its more humanist forms, where bioethics focuses on respect for the patient or the dignity of life, the human being and human (as well as, more recently, animal) life are still perceived as fixed entities, endowed with particular values, on which certain actions should or should not be exercised. It is perhaps not too much of a leap to suggest that humanist bioethics, even if coming from a well-intended desire to protect the weak, often ends up immobilizing the subject it cares about, subjecting it to its own rules and procedures, and, in this way, reinforcing the very hierarchy of power it purports to overturn. Of course, this is by no means to dismiss the politically significant work that has been undertaken under the aegis of bioethics in the areas of patient autonomy or animal protection; it is only to raise questions about broader philosophical assumptions underlying the traditional bioethical models. Turning to Foucault can allow us to revisit the very question of "life" and its constitution. It can also provide us with a new way of thinking about what it means to live an ethical life, and how to care about one's life, without relying on predefined values or legal frameworks.

Life on the Web

June 12, 2007
I think I'm feeling a little off my center. I'm desperately grabbing at spare seconds to do things I want to do: make jewelry, blog, organize my desk, read, sleep . . . and I've been feeling like my life is some sort of bizarre game show where I wake up

and the clock starts and I start running full speed ahead into an arena to do battle with the Predator (that movie scared the shit out of me so hardcore that I am sans shit to this day!) and I am armed with a sponge and some non-toxic cleanser and a list of other things that really really need to get done if I manage to get out alive!!!

And sometimes in that rush, sometimes I move away from the person I have always been.

Advanced Maternal Age
http://advancedmaternalage.blogspot.com/

I would like now to focus on one particular domain of contemporary media culture—blogs and social networking portals such as LiveJournal, Blogger, and MySpace—where millions of users are drawing on a daily basis on what Foucault described as technologies of the self. As explained earlier, I want to suggest that there is something very particular about the techniques of writing, linking, and connecting that are being deployed for the construction of selfhood and agency on- and offline in these Internet practices. Social networking online becomes for me a performative space and practice where what Foucault terms the ethics of the self is being developed and tested. Hence, it is much more than just an example to illustrate a ready-made ethical theory. Instead, blogging and social networking online should be read as *live practices*, where selfhood emerges as always already relational and provisional, where it becomes a project and a task, and where *human* agency is knowingly enacted in conjunction with technology. These practices, I will argue in the further part of the chapter, allow their users to experience the difficulty of, and the simultaneous need for, relationality as a condition of being in the world. The relations established in blogging and social networking are multiple, dynamic, and productive. They bind together and bring forth what I call, after Mark Poster, "humachinic" forms of agency.

Blogging inscribes itself in the horizon of so-called "social media," where user participation, communication, and exchange provide content, and also, to some extent, form, to these Web sites. The aptly titled LiveJournal (LJ) is one of the most popular sites where users keep their blogs. It is also a self-contained virtual community through which people can keep in touch with their "real," "virtual," and "imaginary" friends. Blogger, set up in 1999 but bought by Google in 2003, is another portal which allows users to self-publish their Weblogs as well as "interact with people." LJ and Blogger predate such popular networking Web sites as MySpace (which in April 2007 had 185 million accounts) and Flickr (a photo-sharing Web site which allows photos to be tagged, placed in appropriate user-devised, or "folksonomic," categories, and then browsed by others).[53] The ambition and scope of these online portals should not be underestimated: they aim at creating the experience of "total life" by building intricate systems of connections between online and offline spaces, personae, and narratives. "The updating quality of weblogs is what for [bloggers] makes this text feel 'alive.'"[54] It should be acknowledged that there exist substantial variations in both form and content between different blogs and blog-hosting portals which could undoubtedly be attested to by an ethnographic study of online communities and their participatory practices. Also, I am aware that the "hot sites" of 2008, such as MySpace, Facebook, and Flickr, about which I am writing at the moment, are likely to have lost their appeal or even to have entirely disappeared by the time this book comes out. However, I want to suggest that the call for remaining attentive to the "specificity" of new media as articulated by Katherine Hayles[55] and Mark Poster[56] needs to be accompanied by the increased awareness of what Gary Hall calls its "singularity." Operating in the Heideggerian tradition taken up by Derrida, as well as that of recent work on the study of literature, for Hall this theory of the singularity of new media involves paying far closer attention to the affective, performative aspect of new media and to the singular instances in which online events and encounters take place.[57] As well as recognizing the specificity of MySpace as different from, say, LiveJournal or Bebo,[58] reading new media in terms of their singularity would therefore involve taking heed of the always already different site image that MySpace renders to its users, the singular route the user of the site has taken to get to it at a particular moment in time, the networked trajectory emerging at a particular instant between different users, sites, codes, texts, images, as well as the interpretative endeavors of the media critic, who actively produces these media (as being "this" or "like that") while analyzing them.

Singularity is a way of reminding us that "specificity is not enough," that there is always a remainder—of meaning, of affect—in every new media "event": and it is precisely this remainder as something that escapes even the most experienced user, the most astute media theorist, and the most thorough ethnographer, that also prevents us from reducing singularity to "individual experience" as an identifiable and clearly delimited entity. This is both a practical and a conceptual difficulty that writers on new media have to deal with. Of course, if it is to be understood as a concept, singularity always needs to reach to a certain commonality against which it sets itself: the pragmatic and tentative commonality of categories known as "blogs" or "social networking portals," the shared practices of narrating and linking. My turning to the concept of the singularity of new media is an attempt on my part to explain why I am *not* doing the ethnography of, say, LiveJournal in this chapter or even providing too much "textual analysis" of the blog threads I have woven into this chapter, but also why—while I recognize that there are substantial differences not only between LiveJournal and Blogger, but also between myriad "Blogger events"—I am drawing on the emergent shared understanding of the concept of "blogs and social networking portals" in my discussion of the ethics of the self in the digital age.

The above reservation, even if perhaps too self-reflexive and too circular for some readers, is needed precisely because the singularity and novelty of these new forms of online media are most frequently being analyzed via the frameworks and discourses of old, already known technologies. For some, blogs are online journals, digital age versions of traditional diaries and personal notebooks, where one keeps a regular record of one's thoughts and adventures (or lack thereof), as well as comments on the workings of inner and outer world(s). Others see blogs as a form of independent journalism, where the boundaries between the producers and consumers of news, and between news and commentaries, are being eroded. Initially, blogs contained mainly textual narratives and links to other online texts, but now they also include photographs, video, and audio. What distinguishes them from traditional paper journals is precisely their communitarian and participatory character:[59] blog entries can be linked up to and commented upon by different users (although it is possible to have a private blog to which only invited friends or even no one will have access). Blogs are then seen as a megaphone for the already existing but silenced or little heard voices; they amplify thoughts and passions already in place and help build communities of like-minded people: be it cat-lovers, Java programmers, or anti-globalists. They are also described as a way of empowering "the little guy."[60]

June 11, 2007
We finally get the air conditioning fixed (after a serious argument with Sam the AC guy about how much he charged and a threat over someone getting their legs broken) and you people have the audacity to complain that it's too cold in here?! Are you all out of your effing minds?! The next person who touches the AC controls, and/or covers up an AC vent with cardboard and packaging tape, will be blamed for the downfall of this office—they will be punishing us all! I will call up Sam the AC guy and have him come back and break the damn thing again—and believe me, he will be more than happy to break something of ours!

If you prefer to sweat, do it on your own time.

The Angry Office Manager
http://www.theangryofficemanager.com/

However, some blogs merely serve as amplifiers for the big guys' voices, revisiting the dominant political views and publicizing existent economic enterprises. No matter what their actual content, and whether they are "conformist" or "oppositional," due to the intense practices of free labor they require blogs and social networking are sometimes interpreted in terms of "a full consensus-creating machine," where the work of the digital proletariat can be exploited. This narrative of the exploitation of the individual by the technosystem has been accompanied by another offshoot of the Marxist and, more broadly, modernist critique of popular culture: that concerning the alleged banality of its practices and the narcissism of its participants.[61] As Kris R. Cohen explains, "Bloggers are said to be narcissists because they persist in publicizing their boring lives." However, he also points out that "Both accusations—that blogs are boring and that bloggers are narcissists—rest on standards for public behaviour which never quite make themselves known."[62] Cohen identifies here an interesting contradiction in the public perception of bloggers:

they are seen as occupying a position that is simultaneously too public (i.e., they are too easily noticed by peevish critics, or too easily thought of as pretentious by other bloggers) and not public enough (with the blogger just speaking to himself or herself, or worrying about losing the audience when their hit count drops). What is particularly curious about the position of the blogger audience in these negative accounts is that it "either disappears into a solipsistic void or appears grandiosely exaggerated, subject not to the interests or circumstances of the blogger but . . . to the rhetorical needs of the critic. . . . [T]he absence of a 'real' audience leaves the blogger seeming to stare at herself while thinking she is staring at others—pathetically deluded like Narcissus."[63]

May 30th, 2007
Even though there's a lot going on I'm feeling like I might finally have a rhythm to what I'm doing and almost satisfied with what I'm getting done. Most of all, I'm loving that the opportunities for writing are beginning to match the number and the timing of the ideas I'm having. I am a happy nerdy writer, and in the fourth year of being a postdoc, I guess it's better late than never! It says something, though, doesn't it, about the amount of self-knowledge that's probably necessary when going in to a postdoc, or even a special studies program (that's management-speak for *sabbatical*). That "rhythm" I'm talking about has personal and circumstantial aspects to it that can't be planned.
Home Cooked Theory
http://homecookedtheory.com/

We can see from the above that the narrative structure which frames typical discourses on new media is rather rigid and overdetermined. So how can we get out of this dialectical and, indeed, moralist impasse in theorizing about blogging and social networking online? How, in other words, can we think about these new media practices without immediately valorizing them as good or bad, productive or useless, resistant or oppressive? To do this, I want to develop what might initially seem like a negative thread, which I have picked up from Kris R. Cohen, and entertain for a moment the idea that blogging and "hanging out" on Facebook are perhaps indeed intrinsically linked with narcissism. However, I do not propose to read these practices as symptomatic of the *culture of narcissism*, which is often associated with mass consumption, instant gratification, and constant need for self-renewal, although I do see them as a reinforcement of the ego ideal users construct of and for themselves and then painstakingly attempt to achieve. Even the recognition of personal failures—from "Why did he dump me?" to "Why did my paper get rejected by such and such academic journal?"—seems to provide a way of reinforcing the ego on its way to self-fulfillment and closure. For example, many academic and professional blogs function as veritable "brag spaces," in which one just lists one's achievements: the important people one has met this week, the notable invitations one has received, the publications one has had accepted this year. And yet there is a generous side to both academic and nonacademic blogging, where blogs and other social networking sites operate as a medium through which one "gives" and "shares" interesting notices or funny comments with others.[64] Both of these types of activity—the crafting of the self according to the ego ideal and the sharing of the (data) gifts one has amassed—are narcissistic: the former as a manifestation of one's exaggerated investment in one's image (as a lover, music fan, or professional), and the latter as an attempt to gather the whole world "in my image" and then give it to others. Blogging can therefore perhaps be jokingly described as a delirious activity in which the self attempts to be with others while also recoiling from the wounds that the other (blogger) inflicts on me.[65]

The Media Culture of Narcissism

May 31, 2007
Today is the day where nobody is returning my phone calls or emails. I have left phone messages for at least eight people without being called back. I've also emailed several of these people to tell them to check their messages and to call me back. Eight people, I kid you not. If I have either called or emailed you today, please call or email me back either tonight or early tomorrow, because you are all giving me a massive complex.
Clublife
http://www.standingonthebox.blogspot.com/

Readers may now perhaps wonder whether I have not just come up with another dismissive account of blogging and social networking which inscribes itself in the negative strand of narratives about new media technologies in general, and blogging in particular. But this would assume that we define narcissism as a negative phenomenon, a cultural or personal pathology which has to be overcome for social relations to be established. However, I want to suggest that narcissism is inevitable, or even necessary for sociality, and indeed, to use Derrida's words, that "There is not narcissism and non-narcissism; there are narcissisms that are more or less comprehensive, generous, open, extended."[66] Derrida goes on to argue that what is referred to as non-narcissism is only a more welcoming, hospitable narcissism, one that is much more open to the experience of the other as other. He explains: "I believe that without a movement of narcissistic reappropriation, the relation to the other would be absolutely destroyed, it would be destroyed in advance. The relation to the other—even if it remains asymmetrical, open, without possible reappropriation—must trace a movement of reappropriation in the image of oneself for love to be possible, for example. Love is narcissistic."[67] It is in this sense that an act of reaching to other (through an online posting, a link to someone else's site, or a fantasy of, and desire for, multiple readers) is narcissistic, but also that narcissism is revealed as necessary to establishing this relationship.

Drawing on Derrida's playful account of the ambivalences of our psychic economy, let me thus suggest that this desire for (the death of) the other manifested by bloggers and other online social networkers is actually a condition—even if not a guarantee—of an ethical way of being with others. If we define ethics after Levinas and Derrida as openness to the infinite alterity of the other (leaving aside for a moment the Foucauldian thread developed so far in this chapter), narcissism is revealed as an inextricable part of ethics. There can therefore perhaps be "good" and "bad" narcissism, where "good" and "bad" do not stand here for moral categories sanctioned by predefined philosophical or religious positions but rather for conditions of our subjectivity, our psychic health. The not-so-rare "blogger's delirium," manifesting itself in the constant checking of the site's counter, in comparing herself with other bloggers or in instantiating flame wars in comment boxes—as in Clublife author's moan, "I have left phone messages for at least eight people without being called back"—can be read as an attempt to construct a self. This is a much more serious and difficult project than one focused on merely expressing oneself or even on performing one's identity, whereby one either attempts to convey to oneself and others what one is "truly" like or draws on the set of available props and identitarian positions (a white middle-class female teenager, a suburbanite housewife from Dallas, a dog) to enact a self or play a role.[68] The narcissist's "delirium" is rather like a Nietzschean rapture, a positive condition of the ethos of becoming that we can also find in the work of Foucault.

What is so interesting and promising about Nietzsche's state of rapture, according to Ewa Ziarek, is precisely its reaching to the outside—to what or whom it desires—which "draws the subject out of itself."[69] Ziarek also locates this "attraction of the outside" at the heart of the Foucauldian ethos, "a philosophical life in which the critique of what we are is at one and the same time the historical analysis of the limits that are imposed on us and *an experiment* with the possibility of going beyond them,"[70] an interpretation which offers a possibility of rapprochement between Foucault and Levinas. Foucault himself is less concerned with the ontological conditions of this outside (which ultimately imposes a limitation on his ethics) than with the self's breaking out of the congealed, fixed forms of being in an attempt to imagine and create some new ways of life.[71] However, this drive toward an outside of selfhood as we know it and a desire to go and refashion it already establishes a relation to the outside, to what is not in the self (even if Foucault's own concern will remain with the self's process of becoming rather than with the forms of alterity that make the emergence of this self possible). This rapture or delirium in which the self cannot be contained within its own boundaries can thus perhaps be interpreted as the blogger's enactment of what we have termed "good narcissism," a reaching toward the other (blogger, reader, hacker), the material effects of whose online presence are constantly calling the blogging self into being. This is also a way of countering what Craig Saper describes as b-logocentrism, "a neologism in which the extra b stands for *banal narcissism*, suggest[ing] how blogs can intensify the appearance of a self-present speaker instead [of] a decentered subject in hypertextual webs."[72]

Returning to the dialectical new media narratives sketched out earlier through which blogging and social networking on the Internet are usually analyzed, I want to suggest that to read these practices in terms of the Foucauldian ethics of becoming is not to negate the possibility of interpreting them as political practices in which citizenship is recognized at the microlevel and in which "little guys" are given the voice. But neither is it to subsume these practices under the familiar framework of democracy, political participation, or even friendship, because doing so would mean reducing the ethical possibility of both the self and the technology with which it remains in a dynamic relationship. Nor is it to deny that banality, boredom, and self-obsession constitute part of the experience of blogging or hanging out on MySpace. It is precisely in this tension between many users' sense that these portals are merely mirroring the banality of their own and others' lives and the possibility that they may contribute to the reworking of life forces and establishing a new relation to one's life (on- and offline) that the ethical potential of Live Web lies. Interestingly, Foucault associates the practice of self-writing precisely with an ethos of life. The keeping of individual notebooks and memory books focused on the recollection of the past, or capturing the already-said or what one has managed to hear or read, is for him "a matter of constituting a *logos bioēthikos* for oneself . . . , an ethics quite explicitly oriented by concern for the self toward objectives defined as: withdrawing into oneself, getting in touch with oneself, relying on oneself, benefiting from and enjoying oneself."[73] This phrase *logos bioēthikos* provides a key for my rereading of bioethics as a practice of good life, always on the way to becoming a good life. However, Foucault has in mind something much more material and direct than just a story *about* one's life and how it should be lived: this practice of self-writing is said to produce "a body." Drawing on Seneca, Foucault claims that "writing transforms the thing seen or heard into tissue and blood."[74] From this perspective, diaries, blogs, and online profiles are not just commentaries *on* someone's life, already lived to this point but also somehow more "real" outside its narrative, but rather materializations *of* it. Digital writing and linking is therefore not only a form of *cultural* production but also of *corporeal* one; it literally produces the body by temporarily stabilizing it as a node in the network of forces and relations: between multiple servers and computers, flows of data, users' eyes, fingers and sensations, particles of electricity, and so on.

It has to be emphasized that there is absolutely no guarantee the work on the self, be it in the form of blog postings, MySpace links, or Flickr community activities, will be ethical rather than individualistic or solipsistic. Indeed, Foucault himself recognizes that, even though the care of the self becomes coextensive with life (and it is in this sense that, say, john23's blog is always already john23's life, not just an account of, or a secondary reflection *on*, his life, while his offline activities are somehow more real), this practice of the self which is "theoretically" open to everyone will only be realized by few.[75] Foucault lists the lack of courage, strength, or endurance, the inability to grasp the importance of the task and see it through, and the unwillingness to listen as limitations to the universality of this ethics of the care of the self. It is perhaps not surprising that a great number of blogs and individual profiles on social networking portals present themselves to many media and cultural critics as boring, as banal, as not doing anything at all, and as replicating the most fixed ideas and values (although we cannot, of course, rule out critics' own intellectual and cultural preferences and values reflected in these judgments). However, to say that the care of the self is indeed a "universal practice which can only be heard by a few"[76] and that what we can call "the event of ethics" will therefore be very rare is to assert something else than just that "every site will find its fans" or to argue that only popular sites with the most hits have managed to embrace this ethical potential of the medium. Popularity (or its lack) has nothing to do with the care of the self, on- or offline. I therefore want to suggest something that perhaps goes against the grain of more typical media and cultural studies interpretations of blogging and its users. Rather than seeing LiveJournal or Flickr as being primarily about exchange, with bloggers expecting to be read, responded to, or at least tagged and hence acknowledged in one way or another—"If not, why would they publish their musings on the internet instead of letting them sit in their personal files?"[77]—I propose to read blogging as being as much about experiencing and enacting the simultaneous difficulty and necessity of relationality as a condition of being in the world. This interpretation seems more plausible in the light of the fact that most blogs and online profiles have very few readers: in most cases, the blogger herself.[78] "Social networking" sites function as a testing ground for enacting the dramas and (inevitable) failures of sociality.

June 2nd, 2007
Around 9 we left for the zoo and I immediately got lost in the park. It is staggering to me how stupid I can be. I have been to the zoo a dozen times, the park is clearly signposted, but no matter. I still get lost every time. Once I was out for 2 hours and never found it. I arrived back home half parched to death, Billy screaming with frustration and boredom. Today was a better day. I got lost on a very pleasant nature trail; Billy got to see lots of little critters dashing across our path and numerous waterfalls and streams. It was a beautiful morning, the air was clear, and the temperature was back below 90, thank god.

A Brooklyn lad
http://heidisimon.com/wordpress/

However, even if what we might call "the event of ethics" in the blogosphere is indeed very rare, the enactment of the technology of the self through the techniques of writing and linking (although it has to be mentioned that linking, brought to the fore in online texts, is always already the condition of linear, grammatological writing in the first place) makes the blogosphere a privileged space for analyzing the emergence of the practice of the care of the self. Naturally, we could have explored the enactment of this ethics through the technologies of the self in such cultural practices as dieting, bodybuilding, or reading self-help manuals. There is nevertheless something very particular about how this ethics of the self takes place through writing, especially the writing that involves digital technologies and new media, because of the kind of self and the kind of life that are performed in this process. It could be argued that in blogging and online networking an enactment of a more embodied, aware, and "lively" relationship of the self with technology takes place, and that "life" is thus revealed as always already technological. Foucault himself, when commenting on the coming into vogue of notebooks to be used for personal and administrative purposes in Plato's time, foregrounds the technological aspect of all writing that needs to be taken account of. Providing an interesting gloss to the current debates on the novelty of technology and "new media," he argues that in ancient Greece "This new technology was as disrupting as the introduction of the computer into private life today. It seems to me the question of writing and the self must be posed in terms of the technical and material framework in which it arose."[79]

Social networking sites thus provide an experiential space for actively taking on, rather than merely acting out, the trace of technology in the human self. Recognizing that media users' attention is usually focused on the content of the technology rather than its machinic or formal qualities, Mark Poster states that "Whenever individuals deploy media, they are in the midst of a system of power relations that remains out of phase with their conscious mind."[80] This for him leads to a reconfiguration of the subject–object distinction into what he calls "the humachine."[81] Even if it is the intuitive and unconscious embracing of technology (as this process itself is simplified through the use of templates, etc.) that precedes the self's attempt to connect with others, blogging and social networking return agency to this process of being in the techno-world, of living with technology. This goes some way toward interpreting media culture as a productive apparatus of power, a network of nodes and temporary stabilizations of forces, where the self is neither entirely "free" and autonomous nor entirely and permanently subjugated. Rather, it emerges from this network as technological, or "humachinic," and does so not just through its writing and linking activities but also through the construction of its life—past, memories, dreams, and desires—as technological traces. For Bernard Stiegler, any technical instrument registers and transmits the memory of its use. A carved stone used as a knife, for example, carries a record of an act of cutting and therefore functions as a form of memory. Technology becomes for Stiegler a condition of our relationship to the past, but it also works as a kind of compass, positioning the self in the temporal network from which the linear sequence of events can be distinguished—precisely by the identification of technological traces (cuts on the knife's blade, rearrangements of the online template). Functioning as a support for memory, a technical object marks a particular temporal situation while also recording the passage of time. This is why for Stiegler human beings can experience themselves only through technology.[82]

And it is precisely as the leaving of traces that Stiegler interprets human existence, an existence that is for him always already technological. He argues that what Heidegger calls *Besorgen*, more recently translated as "taking care" (rather than the earlier "concern"), is also "beyond the mere activity that survival requires, the will to *be*, that is to say to exist, to be in the sense of ex-isting, and therefore of *marking, leaving a trace*."[83] The "delete" function on one's keyboard or screen also lets users remove, or at least obscure, their traces, and experiment with ways of narrativizing their life—and death. The MyDeathSpace.com site, which provides an archive of deaths of selected MySpace members (but is not officially affiliated to MySpace),[84] allows for dead members' profiles to be kept "alive" by comments and mementos posted by one's "friends." Of course, at the same time this can all be a hoax: to paraphrase an old adage: on the Internet, no one knows you're a ghost.

The Nodes of Power in the Web of Life

Thursday, November 30, 2006
I am the kind of person who has a very hard time getting over things. My sense of nostolgia [sic] is strong. For instance, it pains me to throw out a receipt from a great meal out with friends, or the ticket stub from an exceptional concert. They're mementos of fun times, and just by looking at them I am brought back to the moment. I like that feeling. However, said receipts and ticket stubs pile up, and really, they're just trash. So, eventually and with a heavy heart, I drop them into the wastebasket (and then, usually, immediately bring the garbage out to the bins so as not to tempt myself to pick them out of the wastebasket). I'm also pretty sure my mother is going to sell the house I grew up in. While dinner receipts are petty, this will be monumental. I may flip out. It won't be pretty. And I don't want to expound on this any more. I'm not quite ready to fully face the inevitable.

Convenient Parking
http://slappar.blogspot.com/

If politics today is indeed focused on the management of the life of both individual citizens and whole populations, then blogging and social networking online emerge as forms of self-governmentality, whereby "individuals are expected to narrate their own lives, connecting their stories more or less closely to preexisting narratives, such as the idea of progress."[85] According to Poster, "Modern society imposes on individuals the task of taking account of themselves, of forming or directing their lives, of intermittently taking stock of where they are at a given point in life's journey, and, at base, knowing their own story."[86] One of the forms through which biopolitics is enacted today is the imperative for online presence: "'If you're not on MySpace,' an American teenager told a researcher, 'you don't exist.'"[87] And yet blogging and social networking also enable the self to establish an active relation to its own life and the processes of its management. It is the taking up of this challenge that *bio*ethics endorses. Bioethics functions here as an ethics of life, whereby the always emergent self takes responsibility not just for its own health but also for life as such or for its being in the world. Becoming and self-creation present themselves as important ethical tasks in this framework. The impetus for this process of self-creation, I would argue—contra Foucault—always comes from the alterity "before me," both in a temporal and spatial sense.

Foucault positions his ethics as response and resistance to the organized forms of power and its historical structuration. Situating it in the context of the relationship with, and pleasure of and for, the other, he defines ethics as "*an experiment* with the possibility of going beyond" the limits imposed upon us.[88] Yet it could be claimed that Foucault's ethics of becoming is unable to account for the source of this possibility of going beyond. In other words, it comes to a halt before the notion of alterity, unable to justify what drives its aesthetic relationship with the other, be it in the form of friendship or pleasure, or explain where the impulse for this opening into the other comes from. For Ziarek Foucauldian ethics limits otherness to the endless variations on the plane of immanence, as a result of which it cannot accommodate an obligation to the other or respond to an external claim.[89] It is in Levinas's ethical call as coming from the always already primordial alterity of the other that we can locate a more convincing ethical impulse, I would suggest. The Levinasian supplement to the ethics of becoming should not, however, be seen as a one-way shift from immanence to transcendence, or as a turning toward God, but rather as a different, pragmatic resolution of the question of ethical injunction, that is, this drive that pushes the self to self-create, to forge life, to become. It is also a way of ensuring that alterity, or difference (seen as commonplace, "banal," and unworthy of serious critical attention by some proponents of an ethics of immanence),[90] does not get reduced to a mere resource for the self.

Bioethics as an ethics of life is therefore situated in, or even arises out of, the tension between bad and good narcissism, between disciplinarity and care, between the Foucauldian *rapport à soi* and the Levinasian response to the alterity of the other, and, last but not least, between self-creation as neoliberal imperative for individualized productivity and an ethical injunction for continuous restlessness and movement. This tension is not a permanent suspension between two sets of equally valid options: it entails a need for a decision, to be taken, always anew, in a singular way, in an undecidable terrain. The possibility of the imperative of the care of the self turning into bad narcissism, b-logocentrism, or "moral dandyism,"[91] and of self-creation becoming a neoliberal project in which the self is seen as the ultimate value that needs protecting, has to be kept in place precisely as a guarantee of the ethicality of this project. Were we to eliminate this possibility (or even danger) in advance, we would be turning our bioethics into a technicized schema for the improvement of the world, predefined and carefully designed by "ethics experts." Technology does nevertheless serve an important role in the ethics of life—not as a threatening other which needs to be overcome for the protection of life or as a set of calculation procedures worked out in advance, but as a container for the tensions between bad and good narcissism, between disciplinarity and care, between self-creation and reaching out to alterity. *Bio*ethics, the way it is emerging in this chapter, is thus always already technological, in the sense that it is predicated upon the acceptance of the technicity of life as its condition of being. It is also a way of allowing us to develop a nonhysterical relationship to technology and a way of living a technical life for the "humachines" that are originarily prosthetic and dependent on their technologies.

To talk about "humachines" is not to posit a seamless one-dimensional flow of life but rather to map out, in a Foucauldian spirit, a network of differentially stabilized, asymmetrical but mutually dependent nodes of power. If biopolitics can be also understood as a way of instituting and managing this network, bioethics is the realm of differentiation within the network, of making decisions about life from within the network—but also of responding to what comes from outside the network, to what transcends its plane of immanence. Thus, attempting to answer the question that frames the last section of this chapter—"What if Foucault had had a blog?"—I conclude that he would have probably been a narcissist, but what I have described here as a good, or ethical, narcissist. As mentioned before, "good" does not stand here for an a priori universal valorization but rather for the recognition of narcissism as an inevitable condition of an ethics of the self, and of the self's being with others.

So, are bloggers narcissistic? Absolutely. But also, inevitably. Are they ethical? Possibly. But, also, perhaps, rarely.

II

Bioethics in Action

4

Of Swans and Ugly Ducklings: Imagining Perfection in Makeover Culture

Only Some Will Make It: Enter *The Swan* (and a Few Ugly Ducklings)

"Four months ago, these nine women were given a once-in-a-lifetime opportunity to change their lives forever. They underwent the most radical transformations, both inside and out. They went from ordinary women to extraordinary beauty queens. Tonight, in the most dramatic pageant in the history of television, one of these lucky women will be crowned *The Swan*."

"This is the most unique experience of our lives."
"Our goal is to transform average women into confident beauties."
"It's a brutal regimen over three months."
"Only some will make it."
"But all will be changed forever."
(*The Swan*, Fox, 2004)[1]

This chapter takes as a starting point a consideration of the extreme makeover show *The Swan*—in which contestants undergo a "total transformation" via radical cosmetic surgery as well as confidence training—within the context of Michel Foucault's and Giorgio Agamben's notion of biopolitics. As we have learned in chapter 3, biopolitics is a form of political regime under which bodies and minds of citizens are administered and under which life is "managed." I want to argue that what is at stake in extreme makeover shows such as *The Swan* is precisely an attempt to exercise such biopolitical domination and to subject the participants' bodies and lives to disciplinary techniques. *The Swan* belongs to the recently popularized genre of "extreme makeover TV," which treats post–*Big Brother* audiences to documentaries featuring the remodeling of real people's homes, gardens, wardrobes, and—as has been the case with shows such as ABC's *Extreme Makeover*, MTV's *I Want a Famous Face*, and Fox's *The Swan*—bodies.[2] There is, however, something singular and unique about the way in which *The Swan* deals with the subject of makeover by framing it in biozoological terms and by introducing the survival of the fittest as its principle of entertainment. The show is designed as a competition among a group of women who are all undergoing a three-month-long "total transformation." This transformation involves cosmetic surgery, a weight loss program, and "personality training," all carried out without the women being able to see themselves in a mirror. Each episode features two competitors who are judged by a panel of experts on the success of their makeover, with the overall winner of the series being crowned "The Swan."

While I position extreme makeover TV as part of the global biopolitics of life management, the aim of this chapter is not merely diagnostic. I am first and foremost interested in the possibility of tracing a counternarrative to this rather gloomy story of biopolitical disciplinarity.[3] It is in the area of "alternative bioethics" that I want to locate this counternarrative. Proposing to read the show's "swans" as twenty-first-century neo-cyborgs bearing the marks of technology on their bodies, I want to explore the promising ethical ambivalence of the kinship between humans, animals, and machines that these bodies exemplify (even if, it may be argued, the show itself ultimately forecloses on this promise).

"Ladies, always remember where you came from and how you got here, and don't forget to live HAPPILY EVER AFTER!"
(*The Swan*, Fox, 2004)

The Biopolitics of Makeover Culture

As explained in chapter 3, biopolitics is a name given by Foucault to political regimes in which power focuses on "the biological existence of a population."[4] Developing Foucault's argument, Giorgio Agamben argues that in modern times the original exclusion of *zoē*, that is, material life or the sheer fact of living, from the polis becomes extended and generalized to the point where it becomes the rule, that is, "the fundamental political structure."[5] In his earlier work Agamben positions the concentration camp and the totalitarian state as the exemplary places of modern biopolitics, but in *State of Exception* he goes so far as to extend the biopolitical framework to the "military order" instituted in the world by the president of the United States in the aftermath of 9/11. I want to suggest in this chapter that radical makeover shows such as *The Swan*, produced in the United States and the United Kingdom and aired on satellite TV stations across the world, constitute yet another example of such exemplary places of modern biopolitics. Even though it might seem imprudent to put *The Swan* in the same category as concentration camps, gulags, or Guantánamo, I am convinced that in the context of the "Iraq makeover" that U.S. and U.K. politicians have orchestrated together with international news stations,[6] the examination of extreme makeover shows such as *The Swan* as applications of the dominant technologies of life management is extremely urgent. And it is the exploration of the "zone of indistinction" between biological and political life that becomes crucial for my parallel reading of the extreme makeovers of the individual bodies of American women and the collective lives of Iraq's population—both accessible to us via TV screens. I should perhaps make a reservation here: I do not argue that individual bodies are "like" states or populations but rather that this equivalence is assumed and enacted by the biopolitical regime. In other words, biopolitics establishes an equivalence between biological and political life, between individual bodies and populations, by applying a similar set of regulating and disciplinary technologies to both.

The link between bodily and political transformation is actually well documented in both European and American political history. Sander L. Gilman, the author of *Making the Body Beautiful: A Cultural History of Aesthetic Surgery*, traces the ideas and fantasies associated with a bodily makeover via cosmetic surgery back to the late-eighteenth-century ideology of autonomy, the belief in the revolutionary potential of the individual and his or her right (or even injunction) to happiness. Speaking about the period following the American and French revolutions, the revolutions of 1848, and the American Civil War, Gilman argues, "The transformation of the Enlightenment notion of self-improvement moved from the battlefield of liberalism to the laboratories and surgical theaters of the later nineteenth century."[7] The reconstruction of the state was perceived as necessitating physical transformation and the creation of new, healthy, and happy bodies—a conviction adopted by the dominant revolutionary movements of the late nineteenth century, such as Zionism, Communism, and Fascism. The visibility of the transformation on the individual level could serve as proof of the broader political change. Gilman writes, "The Enlightenment self-remaking took place in public, and was dependent on being 'seen' by others as transformed."[8]

The "before" and "after" images of cosmetic surgery patients are a legacy of this need for visibility—and so are the photographs and video clips on *The Swan* which constantly compare the participants' "old selves" with their new transformed looks. However, Gilman also points out that the satisfaction resulting from being finally perceived as "normal" is underpinned by the opposite sentiment on the part of the viewers of such a transformation. The viewers take delight in being able to reassert the difference between their own "authenticity" and the makeover participants' phoniness: the latter are only passing themselves off as the "real thing." In the context of this analysis, I want to suggest that the success of *The Swan* as TV entertainment depends precisely on the gap between "normal" viewers and the contestants who want to undergo a transformation. Contrary to a number of TV critics who focus on the identification of both *The Swan*'s audiences and its participants with the American dream—that is, a belief that we can all be happy, beautiful, and successful one day—I claim it is rather the *disidentification* between the two groups that is being achieved here, something that reasserts the viewers' moral superiority and confirms their distance from someone else's physical and emotional pain. It is precisely this disidentification with those in need of a makeover that functions as a hinge between the two layers of contemporary biopolitics *The Swan* embraces—that working on individual bodies and on whole populations. This psychological hinge serves to drum a conservative message home: namely, that it is only others "out there"—overweight "ugly ducklings," depressed women with facial scars and gapped teeth, but also, by perhaps too quick an extension, "diseased immigrants" or "oppressed Iraqis"—who are in need of a makeover. It is they who need the restoration of the signal points of liberalism: freedom, autonomy, and the right to happiness (translated on *The Swan* into "becoming your absolute best"). By isolating the "freaks" on the TV screen for the pleasure and relief of the nation, by making freedom and happiness only skin-deep, the health and well-being of the American (as well as British, Australian, etc.) population is confirmed.

If we follow Gilman's statement that "A concern with 'hygiene' in the broadest sense and aesthetic surgery's role in the physical alteration of the 'ugliness' of the body led the aesthetic surgeon to become the guarantor of the hygiene of the state, the body and the psyche,"[9] we could perhaps go so far as to say that *The Swan* is making post-9/11 America feel beautiful again; it is rebuilding her self-esteem and publicly healing her wounds. As Rachel, the winner of the first *Swan* series, declares in the final beauty pageant, "The reason I would like to be *The Swan* is that I believe that this program has given me so much. It's given me my self-esteem and a sense of who I am, and now I'd actually like to be able to give that back, and teach everybody else that they matter. That they have a place in this world, and they have to believe in themselves and that they can bring out the best that they can possibly be." The public confessions of the final pageant's participants on why they deserve to be crowned The Swan could be seen as exemplifying "the aesthetic and moral sensibility of the new citizen, . . . beautiful and moral and healthy."[10] I should perhaps mention here that I am not attempting to establish a causal relationship between the 9/11 attacks and radical makeover shows such as *The Swan*. I am only suggesting one possible line of interpreting TV makeover culture at a time when human bodies are being represented as increasingly malleable or even disposable, when they are seen as playthings of different political forces, and when tortured bodies, bodies falling down from high buildings to their imminent death, and male bodies sexually abused by females in U.S. military uniforms have become a regular presence in contemporary news media.

I would now like to turn my critical attention to specific techniques of power through which biopolitics operates on the lives and bodies of the population. The ambivalence of Foucault's distinction between the anatomo-politics of the human body and the biopolitics of the population is of particular interest to me when considering the disciplinary technologies of power exerted over the bodies and lives of *The Swan*'s participants (and, by proxy, its audiences). It is perhaps not too far-fetched to say that *The Swan* is consciously evoking comparisons with an army training camp in its description of the regime the contestants undergo for three months after they sign on to the show. The disciplinary procedures the women are subjected to, all planned with military precision by a panel of surgeons and therapists led by the "life coach" Nely Galán (who is also the creator of this show), include nose jobs, eyebrow lifts, lip enhancement, liposuction, collagen injections, dermatological treatment, Lasik eye surgery, breast augmentation, teeth bleaching, the implantation of da Vinci veneers, 1,200-calories-a-day diets, gym training, and, last but not least, "weekly therapy and coaching for confidence and self-esteem." The contestants are supervised on a regular basis—they are being constantly surveilled without the possibility of actually seeing themselves in a mirror for the three months of the "transformation program." The army-general manner in which the show's master and commander Galán gives orders to the contestants and publicly humiliates them for breaking the show's rules (eating a yogurt or, worse, not wanting to exercise enough) playfully employs the tropes of the military regime. Military associations are further drummed home through the choice of contestants: in the first series, an army wife, and in the second, a Texan army captain. The training camp (if we are prepared to call it this) becomes here a zone of exception ruled by martial law, where the lives of the contestants are placed outside of the politico-ethical normativity of the democratic polis. In the camp, their bodies can be cut open, abused, and remolded beyond recognition so that they can be returned to the "normal world," the eschatological space "after the transformation." However, if we pay heed to Agamben's diagnosis that the camp functions as the originary structure of the current world order,[11] *The Swan*'s training center may actually be a visual representation of what is the fundamental political structure today: a space where bodies and lives of *all* citizens are always in the state of exception, where they can be abused, transformed, and discarded at will in the name of transcendent values and through the invocation of a fictitious space "outside"—in which beauty, health, and world peace are to be celebrated (although the program explicitly repudiates any such identification).

Of course, we need to distinguish here between the concentration camp (which lies at the core of Agamben's argument in *Homo Sacer*) and the military camp, and the very different relation to life and death enacted in each of them. If death is seen as an exception in the latter, in the concentration camp the exception becomes the rule. However, it is the troubling structural proximity of these two sites of biopower, and the fact that exclusion and inclusion, as well as life and death, enter the zone of indistinction in a number of different contemporary political formations, that allows me to position the (nondetermined) "camp," after Agamben, as the dominant political structure today. Agamben himself draws on this ambivalence of the (military–prison–concentration) camp, and on its uncanny proximity to the world outside the camp, in *The State of Exception* when he analyzes current international politics practiced by the United States.[12] This kind of philosophical rhetoric is not without problems, though. Andrew Norris argues that this determination that the camp does actually constitute a paradigmatic sovereign structure today is itself a sovereign decision, seemingly instituted beyond the regulation of rule of reason. For Norris, Agamben himself resorts here to a theoretical strategy he imputes to the political sovereign "out there," a gesture that obscures his own performative act as a result of which a particular form of political narrative, with all its negativity, is established.[13] In the light of this analysis, the playful employment of the camp rhetoric and aesthetic by major media production companies—we can mention here Fox's *Boot Camp*, MTV's *Fat Camp*, and NBC's *The Biggest Loser*, alongside Fox's *The Swan*—raises serious concerns about the role of the media in the performative enactment of the ongoing indistinction between exclusion and inclusion, outside and inside, *bios* and *zoē*.

Further questions also need to be asked regarding the constraints on both biological life and political citizenship which are enacted in the realm of the camp. While we should be careful not to collapse the potential vitality of *zoē* with the near-death passivity of "bare life" enclosed in the concentration camp the way Agamben himself perhaps does all too quickly in *Homo Sacer*, we can nevertheless take a cue from him with regard to his analysis of the role and position of marginal, "inclusively excluded" figures in contemporary democracies. Indeed, we could perhaps go so far as to suggest that *The Swan*'s participants are made to perform the role of what Agamben calls *homines sacri*, "holy-cursed" *exceptional* individuals whose lives are devoid of a sacred function in a community. They can be killed but not sacrificed, while their death does not count as homicide.[14] Originally functioning as a limit concept in the Roman social order, for Agamben *homo sacer* is positioned "outside both human and divine law";[15] she or he can undergo abuse which is unpunishable because it functions in the zone of exception. *Homo sacer* thus serves as a constitutive outside to the sociopoliticial order, ensuring its survival and well-being by being banned from it.[16] Significantly, for Agamben it is this ban that constitutes the original political relation. In *The Swan* the ban takes the form of a self-imposed exclusion, which is made evident in preoperative interviews. During these interviews, the candidates explain why they do not belong in the human order and express the desire for joining it. This diagnosis is confirmed by the panel of experts and handed over to the viewers, who are allowed to "see" the unquestioned need for all these normalization procedures the candidates undergo. The bodies and lives of those "ugly ducklings" literally demarcate the borders of the healthy community; they make America (and, by extension, Europe, Australia, "the West") whole. Any ambivalence about the candidate's "abnormal" bodies and lives that they could possibly share with the viewers on the other side of the screen is quickly erased via the triumphalist rhetoric and aesthetic of revelation employed in each episode's finale. When the contestants are placed before the veiled mirror, the success of their transformation depends on the misrecognition they experience when the veil drops and they are faced with their new look. This is the moment when the viewers can rejoice at their own cognitive knowledge for which they have been prepared throughout the show: having killed the *homo sacer*, they are now ready to welcome its resurrected *alter ego* into the healthy community of the living. Like the medieval werewolf, remaining "in the collective unconscious as a hybrid of human and animal,"[17] divided between the state of nature and polis, the ugly duckling can be "charmed out" of its abnormality. As a swan, a "good animal," it can rejoin the dominant political order from which it was previously banned.

To sum up, we can see here that the participants' bodies are being treated as pieces of machinery; they are objects to be owned, manipulated, and symbolically annihilated. However, this anatomo-politics of the human body coexists with the biopolitical management of the population as a whole, including the show's viewers, producers, and participants. What Foucault describes as technologies of the self—"which permit individuals to effect by their own means or with the help of others a certain number of operations on their own bodies and souls, thoughts, conduct, and a way of being, so as to transform themselves in order to attain a certain state of happiness, purity, wisdom, perfection, or immortality"[18]—have become here another form of disciplinary technologies, whereby an act of self-fashioning partakes of the wider regime of biopower that disciplines individual bodies and regulates populations. The contestants' real and symbolic passage from the banned underworld of ugly ducklings to a eugenically driven world of swan beauty in which "only some will make it" is confirmed (or not) by a successful "passing" as a legitimate member of this new community, beautiful enough and transformed enough. The concept of passing, pejoratively understood in the nineteenth century as an attempt to disguise one's real racial self, works well, according to Gilman, in analyzing cosmetic surgery precisely because it foregrounds the racial, eugenic connotations of makeover practices. We can thus conclude that the competition between the two contestants in each episode, culminating in a pageant contest among the winners of all the preceding episodes, reminds us that only the fittest pass successfully and that a total transformation is an impossible dream.

Desire for the Face "before the World Was Made": From Biopolitics to Bioethics

However, it is not only passing on the biological level that is supposed to be achieved in such makeover practices. Gilman links physical transformation with a transcendent desire for mastery and closure:

It is the desire for control, for the face that existed "before the world was made," before we came to recognize that we were thrown into the world, never its master, that lies at the heart of "passing." Mortality is the ultimate proof of this lack of control over the world, but real history, real politics can have much the same effect. Becoming aware that one is marked through one's imagined visibility as ageing, or inferior, or nonerotic, concepts that become interchangeable, can make one long for the solace of that original fantasy of control.[19]

For Gilman the desire for a physical makeover is thus an expression of a deeper fantasy of totality and closure, a yearning for the face that existed "before the world was made." This is also a desire for a world without difference, without the subjectivity of the self that emerges in response to the alterity of the other, and that has to learn how to live with this alterity.

This recognition of, and response to, the alterity of the other is precisely what the philosopher Emmanuel Levinas described as ethics. No matter how many fantasies of my own supremacy, moral superiority, or political power I harbor, for Levinas I always find myself standing before the face of the other, which is both my accusation and a source of my ethical responsibility. An ethical demand is made on me precisely through the face of the other. It is both my and the other's mortality, our being in the world as "being-towards-death," that inscribes our lives in an ethical horizon, and it is the other's death in particular that challenges me and calls for my justification. Levinas writes:

[I]n its mortality, the face before me summons me, calls for me, begs for me, as if the invisible death that must be faced by the Other, pure otherness, separated, in some way, from any whole, were my business. It is as if that invisible death, ignored by the Other, whom it already concerns by the nakedness of its face, were already "regarding" me prior to confronting me, and becoming the death that stares me in the face. The other man's death calls me into question, as if, by my possible future indifference, I had become the accomplice of the death to which the other, who cannot see it, is exposed; and as if, even before vowing myself to him, I had to answer for this death of the other, and to accompany the Other in his moral solitude. The Other becomes my neighbour precisely through the way the face summons me, calls for me, begs for me, and in so doing recalls my responsibility, and calls me into question.[20]

The face Levinas talks about goes against our everyday understanding of this word, as it exceeds the collection of bodily features: "The face is a living presence," "more direct than visible manifestation," "it is expression. . . . The manifestation of the face is already discourse."[21] The face refuses to be contained, comprehended, or encompassed; it cannot be seen, touched, or possessed in any other way by me.[22] What Levinas means by the face can perhaps thus be described as a face "before the world was made"—but we are talking here about *my* world: the face of the other is always already there, waiting for me before I emerge as a subject.[23] Indeed, it is only in relation to difference, to what is not part of me, that my subjectivity will be produced. Naturally, there is no guarantee that I will respond ethically to this "visitation" from the other's face and that I will not attempt to ignore or destroy it. However, my murderous desire, my fantasy of control and mastery, does not change the fact that I am not the source of meaning in this world, that I am just thrown into it.

The desire to possess the face from the time "before the world was made," which Gilman talks about in relation to cosmetic surgery, can be interpreted as a desire for a world without alterity, for the annihilation of difference and return to a fantasy moment when the self was a master of time, space, and language. Even though this fantasy can be said to arise out of a fear of difference and to be driven by racism, sexism, or homophobia, we can also understand it as an attempt to escape from the biopolitical regime that marks some bodies as different—racially, erotically, or in terms of their ability to perform well in the labor market. In other words, we can see it as a psychological defense mechanism that actually incorporates the splinters of the biopolitical thinking it wants to escape from, in the form of racism, sexism, or body- and beauty fascism. Of course, this is not to say that all those who opt for cosmetic surgery are racist or fascist or to decide in advance that having a nose job or liposuction is politically and ethically "wrong." We can explain in emotional or political terms why some people may want to have a hair transplant or teeth veneers done, why they want to be rid of their "ugly" nose,[24] or why Michael Jackson wants to be white—even if each one of these transformations calls for a singular *ethical response*, which is likely to be different when surgery involves tooth correction and when it involves "race correction." I also need to stress that by no means do I want to dismiss cosmetic surgery clients as mere victims of the biopolitical regime, a pitiful object of analysis for a cultural critic who is somehow "above" them.[25] Neither do I want to valorize different cosmetic procedures in advance as "politically or ethically acceptable" (or not). Recognizing that there is no "pure" position when it comes to bodily transformation and that we are all, in one way or another, participating in the culture of body makeover, I am only interested in denaturalizing this desire, as well as raising some questions for the media institutions (e.g., *The Swan*'s producers) that strengthen it. At the same time, drawing on the ethics of alterity, I want to develop a counternarrative to this story of biopolitical hegemony.

The area in which I locate this counternarrative is bioethics. Bioethics, as I am arguing throughout this book, should not be seen as yet another disciplinary practice telling us in advance how our bodies *should* and *should not* be treated. Eschewing the systematic normativity of many traditional forms of bioethics, usually rooted in utilitarianism or universal prescriptivism, my alternative bioethical project arises in this chapter *as a response* to the beautifully monstrous bodies of *The Swan's* participants (and, indeed, its "experts") and to the promising ambivalence of the kinship between humans, animals, and machines these bodies carry—even if, it may be argued, the show itself ultimately forecloses on this promise. If, in the modern state, according to Agamben, life and the body have become biopolitical concepts in which the materiality of life, its biological aspects, and its vegetative functions crisscross our bodies' political and legal roles and positions, an effective bioethics that can counteract the normativity of the biopolitical regime has to be thought through the zones of indistinction between *bios* and *zoē*, matter and concept, human and nonhuman.[26]

It is the interrogation of the last opposition, and of the principles of its constitution, that is the most urgent for Agamben in the thinking of a new politics today. Within this zone of indistinction or indetermination the human functions as "the place of a ceaselessly updated decision in which the caesurae and their rearticulation are always dislocated and displaced anew."[27] I want to argue that this obligation to make a decision, always anew, without merely resorting to fixed genealogical categories, is precisely the source and task of bioethics. Within this ethical framework, the question of the human is inextricably linked with the question of the animal, since it is against the latter that the human is defined as the subject of morality and the agent of politics. Agamben points to the emergence of man "as man" through self-knowledge, a reflexive process which allows man to raise himself above himself and thus become different from himself. The process of differentiation at work in the constitution of the humanist definition of man seems double-edged, as man needs to become different from both the nature and the technology that frame him. *Homo sapiens* thus emerges as "a machine or device for producing the recognition of the human."[28] In this process the nonhuman, the bestial, the technological, and the machinic function as man's conditions of possibility above which he needs to elevate himself.

It could perhaps be argued that Agamben does not give due recognition to the technological element in the emergence of the human: technology still seems rather instrumental in Agamben's own argument (even if it is foundational to his definition of *Homo sapiens*). To explore the role of nonhumans as both constitutive of humanity and a source of active "world making" in their own right, we need to turn to some other thinkers. Donna Haraway and Jacques Derrida have, in different ways, taught us to understand the active being of nonhumans as well as the "originary technicity" of humanity.[29] Indeed, for Derrida the demand that the radical alterity of technicity poses to the human is rerouted precisely through the animal. In a similar vein, Haraway's concept of the cyborg as our technological "other," outlined in her "Cyborg Manifesto," can be read as a productive zone of indistinction between ontological categories, as "[t]he cyborg appears precisely where the boundary between human and animal is transgressed."[30] The concept of the cyborg—a cybernetic organism which hybridizes machine and living organism, "a creature of social reality as well as a creature of fiction"[31]—was introduced by Haraway not only to challenge the imperialist fantasies of the technohumanism of the Star Wars' era but also to interrogate both the constitution of what she terms "natureculture" and the role of technology in setting up what we understand as "the organic." Illicit kinship between the human and the nonhuman has thus always featured highly on Haraway's politico-ethical agenda. Taking a cue from Haraway, I propose to read the "swans" and "ugly ducklings" from Fox's TV show under discussion as twenty-first century neo-cyborgs, bearing the marks of technology on their bodies. The monstrous beauty of the show's participants (as well as its "experts") acts as a testimony to cosmetic surgery in the age of mechanical reproduction, in which the biopolitical distinction is established not between normal and abnormal bodies (as it is in traditional bioethics), but rather between pre- and postoperative ones. Indeed, in the *Swan* universe, it is the transformed bodies that are situated on the side of normality, and it is the most transformed participant that wins the coveted title of the swan (even if, as I argued earlier, the show's success depends on the disidentification between "real" and "transformed" bodies and on the resulting moral elevation of the former over the latter).

The bioethics of human–animal–machine kinship that is emerging here attempts to undo the dominant media practices of biopolitical control *The Swan* explicitly draws on, which are directed at the erasure of alterity. In particular, this bioethics challenges the perception of the animal as a caesura of human–nonhuman separation and reveals the human as always already existing in a prosthetic relationship to its technologies. If, according to Stanley Cavell, how we respond to animals, how we see ourselves standing in relation to them, is a test of how we respond to difference generally, how ready we are to be vulnerable to other embodiments in our knowledge of our own,[32] the use of "the swan" as a framing device for the show requires further analysis. It might be rather tempting to dismiss the use of the swan as just a rhetorical gimmick, a playful reference to a children's tale in which animals serve only as metaphors for human behavior. And yet it is precisely this instrumentality in the use of the swan as a cultural concept representing radical visual transformation that deserves our attention. It is also the proximity between the animality of the swan and the imperfect humanity of the female contestants that is of particular interest to me. Earlier on I suggested that the show's participants, under the guise of ugly ducklings, were made to perform the role of *homines sacri*, people whose lives were devoid of a sacred function in a community and who could be killed, mutilated, and abused, but not sacrificed. However, the sacrificial role of the ducklings and swans is not insignificant either. Cary Wolfe, author of *Animal Rites: American Culture, the Discourse of Species, and Posthumanist Theory*, argues that "the full transcendence of the 'human' requires the sacrifice of the 'animal' and the animalistic, which in turn makes possible a symbolic economy in which we can engage in a 'noncriminal putting to death' (as Derrida puts it) not only of animals, but other *humans* as well by marking them as animal."[33] We should understand by now that we are not talking about a straightforward sacrifice but rather a disavowed one, in which both the act and the sacrificed object are reduced in significance. The ultimate disavowal seems to belong to the animal (of the bird variety).

The Swan can thus be said to be enacting the sacrificial economy of our culture, which structures the humanist idea of the human. The logic of this economy is precisely one of disavowed sacrifice: it is not based on a simple substitution through which animals would be killed *instead of* humans. Wolfe observes that, as "we do indeed kill humans all the time . . . it is in order to mark such killings as either 'criminal' or 'noncriminal' that the discourse of animality becomes so crucial."[34] The use of the allegedly innocent "concept" of the swan (and the evocation of the never actually named ugly duckling) is exemplary of the ideological work of othering not only animals but also other humans that do not conform to a biopolitical idea(l) of humanity—in this case, those with leaky, disabled, obese bodies, crooked teeth, and racially suspicious noses. The female participants on the show are temporarily objectified precisely by being reduced to "meat," by being dismembered, cut open, remolded—all with a view to achieving eventual cultivation.[35] Indeed, it is only after undergoing husbandry at the hands of the experts that the participants' bodies are sexualized: at the minipageants at the end of each episode disposable (nonhuman) meat is transformed into desirable (female) flesh. Sexism is thus revealed as a flip-side of speciesism.[36] Both of these function as structuring conditions of the biopolitical logic of modernity, which sees the bodies and lives of others—fat women with crooked teeth, not-yet-democratic-enough Iraqis—as always already in need of a makeover.

However, even if *The Swan* (unknowingly) reveals this logic of disavowed sacrifice as a structuring device of modern biopolitics, in which the value of the human depends on the sacrifice of both women and animals, the show itself forecloses on any further interrogation of this logic. Culminating in the sacrificial pageant in which animal (swan) is both replaced and enacted by woman (at the start of the final episode, each of the pageant's participants arranges her body into a swan-like figure) and in which woman is replaced by a mechanically reproduced "cyborg" version of herself, the show denies any kinship between animals, humans, and machines. It thus withholds the possibility of a nonhumanist bioethics.

The use of "the swan" as a framing device for the show cannot be easily overlooked. Indeed, the question of the animal is fundamental to any enquiry into culture, politics, and ethics today because of the animal's role as a "constitutive outside" in the dominant Western conceptions of moral and political philosophy. The generic animality of the animal has served as a fault line against which the humanity and superiority of the human—including the ability of humans to order the world according to their own categories of preference and pleasure—have been ascertained. It is through Donna Haraway's encounter with animals in her 2003 *The Companion Species Manifesto: Dogs, People, and Significant Otherness* that I would like to explore the possibility of opening up this generic concept. Haraway's book provides a useful starting point for thinking about a new bioethics which embraces the kinship of humans, animals, and machines through the notion of "companion species." *The Companion Species Manifesto* can be seen as an update on Foucault's analysis of biopolitics in its foregrounding of the coexistence of different species through technologies of biopower, and its working out of a bioethics of significant otherness which raises questions for the superiority of the human in the ecosystem. Initially it might seem strange that the author of the celebrated "Cyborg Manifesto" and an astute critic of technoscience should turn her attention to animals in this 2003 pamphlet. However, Haraway is a zoology graduate who has done extensive research on primates.[37] As explained earlier, animals were already of interest to Haraway in her celebrated "Cyborg Manifesto" from two decades ago.

They become even more important in *The Companion Species Manifesto*, but this does not mean that there is no room for cyborgs in Haraway's later argument. Cyborgs join here a bigger family of "companion species," a concept she finds more useful in guiding us "through the thickets of technobiopolitics in the Third Millennium of the Current Era."[38] In the age of soft technologies leading to the development of the patented, cancer-bearing OncoMouse or the first cloned animal, "Dolly the sheep," it is perhaps to be expected that Haraway's dogs are cyborgs of sorts, another example of "category deviants" inhabiting the complex networks of the technoscientific world, in which life is manufactured and nature is technological.[39] The singular, experiential materiality of dogs is of particular importance to Haraway, as dogs "are not here just to think with. They are here to live with."[40] It is precisely the singularity of dogs, evidenced in her moving stories of writer J. R. Ackerley's love for his German shepherd bitch Tulip, or Haraway's godson Marco's training with the family dog Cayenne, that opens up the generic category of the animal. It is also from this singularity that Haraway develops her "bottom-up" theory of ethics in this manifesto. Her earlier texts, such as *Simians, Cyborgs and Women* and *Modest_Witness@Second_Millennium*, were invaluable in tracing connections between bioethics as a neo-Darwinian philosophy that brings together organic and social processes, on the one hand, and the corporate structures of the biotech industry on the other. However, it is only in *The Companion Species Manifesto* that Haraway actually puts forward some more specific ethical pointers. The origins of her ethics of companion species, developed on the basis of "many actual occasions," are experiential and lie in "taking dog–human relationships seriously."[41] Relationality is crucial to this ethics of companion species, an ethics based on an ontology of coevolution and co-emergence between humans and dogs, in which "none of the partners pre-exist the relating."[42] It is thus important to emphasize that it is not dogs that constitute a companion species *for us*, but rather "We are, constitutively, companion species. We make each other up, in the flesh."[43]

Haraway's ethics is hybrid in its origins: it draws on the phenomenological theory of embodiment, A. N. Whitehead's concept of the prehension of actual relations, as well as Bruno Latour's actor–network theory, here translated into a contingent foundation of "multidirectional flow of bodies and values."[44] Her objective is to interrogate the heterogeneous cohabitation of people and dogs, in specific natural territories which are always already shaped by technology, without rooting this interrogation only in the desires and needs of "man." It is also to envisage ways of "living well together with the host of species with whom human beings emerge on this planet at every scale of time, body, and space."[45] Haraway explains: "I believe that all ethical relating, within or between species, is knit from the silk-string thread of ongoing alertness to otherness-in-relation. We are not one, and being depends on getting on together."[46] It is love that functions as an ethical bind between the companion species—although Haraway is careful to distinguish it from technophiliac or caninophiliac narcissism, that is, the belief that dogs are either "tools" for human activity or sources of unconditional love and hence spiritual fulfillment for humans. Instead, love names the ethical co-emergence and cohabitation between specific, historically situated dogs and humans. Love does not therefore mean a mere intensification of affect, or a more proprietary form of possession (as in the phrase, "I love animals"), but rather a preparedness to examine the interrelations between humans and other species and to enter the kennel in order to "listen to the dogs."

Still, this idea is not without problems. Even though Haraway rejects a humanist standpoint in her ethical theory and opposes the reduction or approximation of nonhuman companion species to humans, the values she promotes as crucial to her ethics of companion species—love, respect, happiness, and achievement—have a distinctly human "feel" to them. Indeed, it is the human who defines the meaning of these values and their appropriateness for all companion species. It should be acknowledged that Haraway goes to great trouble to ensure that the needs of dogs are respected and that the understanding of dogs develops from listening to them, from observing their bodies and behavior. She herself talks to those experienced in working with dogs, such as animal trainer Vicky Hearne, to provide a more thorough account of these "needs." There is no escape, however, from the philosophical quandary that even the most committed effort to give the dogs what they want, not what humans want for them, inevitably depends on the human ideas of "want," "satisfaction," and "gift." Drawing on the environmental feminist Chris Cuomo's "ethics of flourishing" does not get Haraway off the hook of humanism she is so keen to avoid either. This is only confirmed in her embracing of Hearne's idea that dogs need to be seen as "beings with a species-specific capacity for moral understanding and serious achievement,"[47] whereby the ethics of companion species becomes, disappointingly, a mere extension of the moral standpoint rooted in the human as a rational, self-reflexive agent.

This is not to say that the (posthuman) *bio*ethics of companion species is impossible, or, more absurdly, that dogs should tell "us" what "they" want, only that a value-driven theory of the good is not the most appropriate basis for this kind of ethics.[48] At best, Haraway sounds like a well-meaning Habermasian who believes companion species such as humans and dogs can work out together, in the process of joint "deliberation," a mutually satisfactory strategy for coexistence, at worst—for example, when she calls for "agility" as a "good in itself" which allows both sides "to become more alert to the demands of significant otherness"[49]—as a good eugenicist. The narratives Haraway tells us about dogs are interesting and passionate: she teaches us how to engage seriously with other companion species, without resorting to either sentimentality or speciesism. Her analysis of the technoscientific apparatuses at work in the management of the lives of humans and animals—such as her account of the "Save a Sato" foundation, which rescues stray dogs in Puerto Rico and prepares them for adoption in the United States—raises important questions about the biopolitical character of Western democracies. However, as in her previous work, Haraway fails to provide a convincing theory of ethics for different species and kinds. What therefore starts as a radical enquiry into the conditions of interspecies coexistence ends up like a recipe for *liking animals a lot.*[50]

Even if Haraway's own "unrepudiated" (and unacknowledged) humanism or her unaccounted-for normativity disappoint somewhat, *The Companion Species Manifesto* is an important text when it comes to thinking about bioethics otherwise. It allows us to envisage a novel relationship between humans, animals, and machines, foregrounding technology as the formative force in the structuring of the "naturecultural" relationships between species. It also raises questions for the status of the swan as a conceptual ornament in a makeover show but also leads us to challenge the erasure of technologies—medical, broadcast, or reproductive ones—in the emergence of "swan beauty" (where the participants are seen as finding their "true self"). By saying this I am not trying to reclaim agency for technology here, I am only pointing out that its placement in the service of (wo)man and her beauty reduces the complex network of connections between human and nonhuman bodies, relations, processes, and practices. It can perhaps be mentioned in passing that it is precisely such a perception of both technology and the animal as being in the service of "man" that has added to the decimation of actual swans in our habitat. As the Web site for the Swan Sanctuary in Shepperton, Middlesex, United Kingdom, explains: "In addition to the natural threats [swans] face from foxes, mink & botulism, modern society has added several more such as pollution, vandalism, uncontrolled dogs, fishing-tackle and lead poisoning, as well as unmarked pylons, overhead cables & bridges."[51]

Haraway's rigid critique of the biopolitical economy of our educational and business institutions which research, control, and subsequently own life, explored most dramatically in her study of the development and patenting of OncoMouse by Harvard and DuPont laboratories in *Modest_Witness*, raises bigger questions about the ownership of human and nonhuman bodies and their role in the flows of multinational capital. To return to our example, it is not only the swan that is transformed from kind to brand in the show: the participants themselves, sporting da Vinci tooth veneers and undergoing, among others, a Zoom bleaching and Lasik eye surgery, become brands—commercial, corporately produced and owned goods, stamped with identical trademarked beauty procedures. (The winner of the Swan prize will receive a $100,000 contract as a spokesperson for NutriSystem.) They become nodes in the flows of capital, connecting beauty, diet and cosmetic surgery industries, global media programming, and Hollywood makeover myths. Borrowing ideas from Haraway, we can develop a counternarrative to the story of the "incredible transformation" told by the show's producers and presenters if we examine what happens to both swan and woman in this transformation of the participants into mechanically reproduced versions of themselves (and of millions of other women subject to the Hollywood beauty regime). Developing this counternarrative is one way of ensuring we stop erasing animals from the story of the technocapitalist optimization of life, as in the computer-driven jargon of mice, bugs, and spiders' webs, whereby animals only function as metaphors for the disembodied world of technology, things to either use for our own pleasure and benefit or kill in order to increase our comfort.

Thinking in terms of companion species which are both coexisting and co-emergent may deprive us of a fixed foundation for our new bioethics, but it does not absolve me of a responsibility to develop it. (This responsibility, I hasten to add, is "mine" rather than just simply "human," hence my vacillation here between the plural and singular pronouns, between "us" and "me"). To sum up, the bioethics that is emerging here as a counternarrative to *The Swan*'s biopolitical hegemony distinguishes itself from traditional normative bioethics on a number of counts. It does not consist of a set of rules on how to treat human and nonhuman bodies and lives but rather of a content-free obligation that these other bodies and lives make on me, and that call on me to respond to them. The ethical response would consist in the minimization of violence, it would be a form of hospitality toward alterity that responsibly negotiates, always anew, between the self's desire for sovereignty and self-sufficiency and the other's challenge to this sovereignty. This bioethical hospitality differs from the "no compromise" posture of deep ecology, so astutely critiqued by Tim Luke and Cary Wolfe for attributing human qualities, and giving at least somewhat human status, to the nonhuman realm of nature.[52] It does not therefore amount to awarding all difference in all life forms—humans, sheep, salmonella, anthrax, and cholera microbes—equal value or promoting biodiversity as inherently and undisputedly good (as a consequence of which we would need to value a priori rare and endangered species, e.g., a California condor hatchling, over more abundant ones, e.g., a human child).[53] Instead, bioethical hospitality as I understand it consists in making a "minimally violent" decision, always anew, under the conditions of impurity in which the defining concepts and material conditions of lives are always already implicated in those of other species and life forms. (It is on the level of "soft" and "hard" technologies—tools, genetic inter- and cross-breeding, machinic and digital reproduction, farming, immunization, computing, bioinformatics, patenting, flows of capital, language—that these co-implications are constituted.)

As explained earlier on in the book, the appropriateness of Levinas's philosophy for devising a nonhumanist ethics has been contested by a number of thinkers, given Levinas's dismissal of the animal as too stupid to have ethics, possessing no reason, and being unable to universalize its maxim. Wolfe aptly summarizes Levinas's position as proclaiming that "the animal has no face; it cannot be an other."[54] And yet I want to suggest that Derrida's ongoing and passionate engagement with Levinas—starting from his early essay, "Violence and Metaphysics," in which Derrida puts in question the possibility of there being an "absolute" alterity of the other, through to the book arising out of his funeral oration to Levinas, *Adieu,* and his 1997 Cerisy paper, "The Animal That Therefore I Am (More to Follow) . . ."—provides us with a way of thinking a nonhumanist ethics of alterity rooted precisely in the demand of incalculable difference which has to be responded to always anew. While in Levinas "the alterity of the other is once again hypostasized (as 'man') rather than left open (to the possibility of the nonhuman other), so that the 'incalculable' essence of the ethical relationship turns out to be not so incalculable after all,"[55] in Derrida this alterity is radicalized by inhering the prospect of an uncertain, perhaps monstrous arrival of the other for whom we do not yet have a name or concept, but also by considering that the radical, non- or inhuman alterity of the other is perhaps a (disavowed) part of what we designate as "human" rather than being diametrically opposed to it. Derrida does not instruct us that animals, cyborgs, or machines are *like* humans but rather that all these identitarian categories emerge only through fixing alterity as being always already "on the outside" of the one that is currently being defined. Through this process, the "animal" has become a "catch-all concept," naming "*all the living things* that man does not recognize as his fellows, his neighbors, or his brothers."[56] It is "the word men have given themselves the right to give."[57] This operation requires a prior fixing of the absolute difference of the human, a difference founded upon what is nonhuman and thus making the animal (hu)man's condition of possibility.

The ethical impulse that emerges from Levinas and Derrida does not lead to the development of prescriptions on how to treat human and nonhuman life forms, but it does highlight the responsibility on the part of those who have designated themselves as human—on the basis of their rationality, sensibility, and mastery of language—to put these "human" characteristics to good use and respond responsibly to the current structuration of the world. It is therefore in Levinas's thinking about ethics as unconditional demand on my being by the alterity of the other that I see a rejoinder to Haraway's normative bioethics, and it is in Derrida that I trace a supplement to Levinas's own humanism. It should be evident by now that bioethics, the way I understand it throughout this book, is not just concerned with human life, and that it does not merely extend the principle of life (and the requirement for its protection) to animals. Instead it raises questions about the animal as a border concept against which the distinction between human and nonhuman (including the machinic and the technological) is made. Importantly, this alternative bioethics of kinship between humans, animals, and machines deals not just with biological life (*zoē*) but also with political life (*bios*), while also recognizing, pace Agamben, the vitality and robustness of *zoē*.

In the context of the ethical issues concerning cosmetic surgery, body modification, and women's beauty regimes explored in this chapter, I realize that there may be something rather frustrating about a bioethics that refuses to evaluate the morality of the actions in which the producers, participants, and audiences of the radical makeover show *The Swan* are engaged. And indeed, the kind of bioethics that is emerging here will not provide us with a definite set of values concerning cosmetic surgery; it will not teach us where to draw the line between necessary and cosmetic procedures. Neither will it help us decide in advance whether people should or should not engage in beauty transformation or whether the correction of a bumpy nose is more justifiable morally than breast augmentation or having an ear implanted on one's arm.[58] What it will do instead (perhaps adding to the frustration of those who already have clear expectations of the tasks bioethics should undertake and the questions it should answer) is shift the parameters of the ethical debate: from an individualistic problem-based moral paradigm in which rules can be rationally, strategically worked out on the basis of a previously agreed principle—for example, that it is the sum total of happiness of all beings that counts or that I should respect the (human or even nonhuman) other as much as I love myself—to a broader political context in which individual decisions are always involved in complex relations of power, economy, and ideology. It is precisely out of this tension, between the need to respond to the alterity of the other always in a singular way and the fact that there is more than one other in the world that is simultaneously making a demand on me, that a responsible nonfoundational bioethics can emerge.

5

The Secret of Life: Bioethics between Corporeal and Corporate Obligations

Cracking the Secret of Life

This chapter starts from the premise that, if there is something like an ethico-political imperative in the humanities and social sciences that underpins their quest for knowledge[1]—that is, an imperative to respond to, and take responsibility for, incalculable difference in what we term "culture" and "nature"—this imperative obliges us to address the most "vital" issue through which difference manifests itself: the issue of "life." Drawing on a number of rhetorical tropes that have played a significant role in the life sciences, I want to suggest that cultural critics have an obligation to engage with the tropes of life developed both in science and in popular discourses about science. Specifically, I propose to look at the conceptualization of the discovery of the structure of deoxyribonucleic acid (DNA), and of the mapping of the genetic code,[2] through the trope of "cracking the secret of life," a trope which has to a large extent shaped the dominant ideas about life, nature, and the human. It is not my intention, however, to position cultural criticism as a corrective to science, a superior critical discourse that can be used to adjust scientific error, as I am aware of the complexity of the debates around questions of life and vitality conducted within the sciences themselves, as well as the ongoing discussions among biologists, social scientists, and cultural theorists regarding the ideological, representational, and performative aspects of the trope of secrecy in the narratives about life.[3]

My own contribution to these debates, as shown throughout the book, lies in the area of ethics. Through an engagement with the discourses of the life sciences, with their rhetoric and materiality, I want to continue in this chapter with my exploration of the possibility of outlining an "ethics of life," a new way of thinking about *bios* informed by the parallel trajectories of continental philosophy and media and cultural studies. This alternative bioethical framework is intended to go beyond some of the more established ways of thinking about bioethics that have been developed within the corporate world of biotechnosciences, as well as beyond the dominant positions on life within traditional moral philosophy. Although, as demonstrated in chapter 1, bioethics today is rooted in diverse philosophical positions, they all presuppose a certain idea of good, a rational human subject that can make a decision about this good, and, last but not least, a need to universalize and apply a moral judgment to a particular situation. It is these three aspects of dominant moral philosophy—predefined normativity, human subjectivity, and universal applicability—that I want to challenge in my questioning of what I refer to as "traditional bioethics." What is of principal interest and importance to me in this chapter is thus not so much cracking "the secret of life" once and for all but rather exploring the already existing cracks in the current discourses and debates on *bios* and bioethics, as well as considering an emergence of a different bioethical proposition from within these cracks.

It should be clear by now that when I am speaking about "the secret of life," I am not referring to general, nebulous formulations that capture the unknowns concerning human life on earth in a more metaphysical or abstract sense: our origin, destiny, quest for the meaning of life, and our unpreparedness for death. Instead, "the secret of life" trope is connected here with a particular moment in the history of the natural sciences, when life was redescribed as a secret that needed cracking and a mathematical problem to be solved, with the solution already looming on the horizon. As the crystallographer J. D. Bernal put it in 1967, "Life is beginning to cease to be a mystery and becoming . . . a cryptogram, a puzzle, a code."[4] It was the very scientists involved in the mapping of the biological processes at the cell level—and, more specifically, in the analysis of DNA in the newly emergent conceptual entity which became known as "the gene"[5]—that drew on the rhetoric of secrecy in order to convey the significance of their research to the wider public. Inspired by the 1944 book *What Is Life?*, penned by the founder of wave mechanics, Erwin Schroedinger, numerous physicists took up this eponymous question in the 1940s and 1950s as a new vector for their scientific careers. It allowed them to leave behind allegedly more ambitious physics, which had nevertheless become tainted in the aftermath of the atomic bomb experiments, and take up biology—previously seen as a "softer" and easier option—while at the same time borrowing physics' authority and methodology. Consequently, in the middle of the twentieth century biology underwent a process of conceptual recalibration—from a science where the language of mystery had its functional place to a science more like physics, "predicated on the conviction that the mysteries of life were there to be unraveled, a science that tolerated no secrets."[6] In other words, the secret already implied the need to crack it, and a realistic possibility of achieving this.

As the author of *The Double Helix*, James Watson (who, incidentally, started his academic career in biology, although he had always been actively interested in physics), acknowledges, Schroedinger's argument that life could be thought of in terms of storing and passing on biological information resonated particularly well with his own dislike of vitalism, a belief in the mysterious forces of life "emanating from an all-powerful god." Watson admits to being totally swayed by "the notion that life might be perpetuated by means of an instruction book inscribed in a secret code."[7] Even though Schroedinger's book belonged to an earlier era in biology, one based on older ideas of cellular organization, and focused on permutations in proteins rather than on relations between proteins and DNA (as would be the case in molecular biology),[8] it paved the ground for a rhetorical shift in the life sciences toward the perception of life as information, before the discipline as a whole underwent a paradigm shift from protein- to DNA-based explanations of heredity.[9]

Defining life, or, more specifically, the genome[10]—although the exchangeability of these two concepts is precisely what concerns me in this chapter—as an information system, a message written in DNA code which needed cracking, was part of a broader epistemic transformation taking place in the 1950s, when cybernetics, information theory, and mathematical theory of communication made a substantial impact on a number of disciplines. In *Who Wrote the Book of Life? A History of the Genetic Code*, Lily E. Kay traces the development of cybernetics and information theory in the context of academic and military work on secrecy systems and cryptanalytic techniques in post-World War II America. She explores this shift toward the discourse of information as a dominant paradigm in the life sciences, where life is no longer seen as a mystery but rather as a secret to be deciphered, via an account of the emergence of communication theory in the United States. Kay explains that Claude Shannon's 1948 paper, "Communication Theory of Secrecy Systems," treated cryptology in information-theoretical terms, with information being seen as quantifiable, transmissible over different media, and bearing no semiotic value (i.e., having no "meaning"). Shannon's general theory of communication, developed from his earlier work on telegraph transmission, "seemed to generate a theory applicable, in principle, to any system, physical or biological, in which information can be properly coded, quantified, and manipulated through time and space."[11] Although Shannon himself, as Kay notices, was against the extrapolation of information and communication theory to genetics, his work on the transmission of information, combined with Norbert Wiener's redescription of heredity in terms of message and noise, led to numerous attempts by other scientists to trace parallels between machinic and organic systems, between automata and cells.

Biology's alliance with the information sciences via a shared interest in "code" in the mid-twentieth century, coupled with a promise of a search for an answer to the "vital" question of life, strengthened its authority and significance even further, as explained by the feminist critic of science, Evelyn Fox Keller. The specific scientific task of attempting to map the structure of the essential "carrier of life," DNA, was given extra valence through the cunning use of a figure of speech known as metonymy: a series of laboratory experiments and modeling exercises *became equivalent to* cracking the secret of life. Seeing these experiments as leading to the discovery of nothing less than "life itself" was indeed necessary if the project was to be taken seriously, both by the funding bodies and the cultural and media agencies that quickly transmitted the mission to the public. The mapping of the DNA structure thus began to stand for the cracking of the secret of life. The story goes that the modeling of the double helix structure of DNA as consisting of base pairs made up of four chemicals—adenine, thymine, cytosine, and guanine—led to Watson's collaborator Crick proudly announcing to everyone in the Eagle pub next to their laboratory in Cambridge on February 28, 1953, that they had found "the secret of life."[12] It was this announcement perhaps that gave impetus to the further proliferation of this kind of reductionist discourse in the popular imagination. This newly developed discourse on life focused exclusively on singular, if imaginary, entities—genes—at the expense of the processes occurring between multiple genes and proteins, not to mention environmental influences or social and political causes, which also played a significant role "in life."[13] The "secret of life" trope was quickly picked up by other scientists as well as the media in their reports of the "discovery" and its consequences. In 1987 the U.S. network PBS made a series of episodes entitled "The Secret of Life" for their educational TV program NOVA, to be followed by the 2001 series "Cracking the Code of Life."

To sum up, the shift to the information paradigm in biology seems to have been to a large extent a strategic *rhetorical* gesture, intended to performatively align the allegedly less prestigious academic discipline with the more powerful fields of enquiry, such as physics and computing, which at the time were receiving sponsorship from government military programs. It was also an attempt to capture public imagination through the daunting metaphysical connotations that "cracking the secret of life" entailed. This explains why, even if most scientists did not "believe" in this rhetoric, they continued using it. Indeed, Kay points out that borrowing the concepts of code and information for biology was first of all a rhetorical maneuver. "The code of life" or even "genetic code" were nothing else but metaphors, as, from linguistic and cryptanalytic standpoints, the genetic code is only a table of correlations, not a relationship between two distinct linguistic systems that would involve changing one set of signs into another.[14] She also explains that, unlike in machine communication, information in biology cannot be easily measured, nor can the materiality of its channel and its genomic, cellular, organismic, or environmental context be ignored. In contradistinction to technologically specific communication theory as developed by Shannon and Wiener, the semantics of information in biology *does* matter and cannot be reduced to mere quantity.[15] Genetic information is content specific; it is not a meaning-free message. Nevertheless, the technical inconsistencies did not prevent the radical rhetorical shift to information discourse, as Kay points out: "despite the acknowledged technical impotence of information theory in molecular biology, its discursive potency intensified by compromising its technical structures. . . . The discourse of information linked biology to other postwar discourses of automated communication systems as a way of conceptualizing and managing nature and society. And it provided discursive, epistemic, and, occasionally, technical frameworks for the scriptural representations of genetic code."[16]

This rhetoric of coding and secrecy is still quite prominent in popular representations of genetics. To celebrate the fiftieth anniversary of the mapping of the double helix, on February 17, 2003, *Time* magazine published a special issue, with the cover story by Nancy Gibbs, "The Secret of Life," claiming that "Cracking the DNA code has changed how we live, heal, eat and imagine the future." Gibbs writes further: "Any 4-year-old who likes ladybugs and lightning can tell you that life is wildly beautiful as far as the eye can see. But it took the geniuses of our time to reveal how beautifully ordered life is deep down where we can't see it all—in the molecular workshop where we become who we are."[17] Her piece is preceded by an article by Michael D. Lemonick, who promises to explain to us how "Two unknown scientists solved the secret of life in a few weeks of frenzied inspiration in 1953." On its Web site the pharmaceutical company Pfizer makes it clear it sees itself as part of the new caste of code masters through a comically ambiguous title: "Genome: the Secret of How Life Works, made possible by Pfizer."[18] To coincide with the golden anniversary of his and Crick's discovery, in 2003 Watson penned another book, appropriately titled *DNA: The Secret of Life*. Last but not least, the December 2007 issue of *Wired* carried an article on the new Silicon Valley startup called 23andMe, a company offering users a subscription to an online service which provides individual and family genetic profiles across generations. The significance of the project was highlighted via a bold slogan on the magazine's cover: "Your life decoded."[19]

Significantly, the development of the "secret of life" project was accompanied by a thorough redefinition of what counted as "life." Poetic, philosophical, and religious concepts aside, molecular biologists moved away from the traditional description of life used by the natural sciences in terms of growth, development, and reproduction to its definition as instructions or "information" encoded in the genes, or, more simply, a code or code-script.[20] As Watson puts it in the Introduction to *DNA: The Secret of Life*, "The double helix is an elegant structure, but its message is downright prosaic: life is simply a matter of chemistry."[21] These pronouncements did not make life any less mysterious, though. According to Dorothy Nelkin and M. Susan Lindee, DNA itself was quickly recoded in popular discourses as "a sacred text that can explain the natural and moral order," "The Bible," the "Book of Man," and the "Holy Grail." Giving mystical powers to a molecular structure in popular, or rather "general," culture led to the perception of DNA as autonomous and independent of the body. Nelkin and Lindee explain that modern molecular genetics promised "a 'complete' understanding of human life, but such promised knowledge, in the form of genetic engineering and genetic therapy, also commonly appears as dangerous and taboo."[22]

However, the definition and redefinition of life in the life sciences, or the way scientific developments have affected the understanding of life in popular knowledges and discourses, are not the primary focus of this chapter, which is why I am unable to do justice to the complexities of the debate on the issue. As suggested earlier, many scientists were suspicious about positioning life as a secret to be cracked. Even if they did use the "secret" trope, they did so strategically and rhetorically—although from a humanities scholar's point of view such a prevalent rhetorical use of a phrase cannot be left unexamined, even if the user is aware of it being "only" a figure of speech. But what I am first of all interested in here is *the specific historical conjuncture of life and secrecy*, when life was redefined as a task to be solved, a code, or secret to be cracked. Aware of the fact that there were doubts about this formulation in the scientific community already at the moment of its pronouncement,[23] and that the contestation of the notion of code had been ongoing among biologists themselves, I want to explore the political mechanisms of revelation and occlusion that were at work in the formulation of these tropes, tropes that have managed to exert so much force on public imagination for years to come. This question of the veiling and unveiling of the secret of life therefore presents itself to me as a political and ethical question; it is a question of taking responsibility for, and a decision in and about, language, and bearing the political and moral consequences of this decision.

The link between "the secret of life" and politics was already implicitly established by the physicist J. D. Bernal in his pronouncement about life ceasing to be a mystery and becoming only a "puzzle" or a "code," which I evoked earlier. This statement implies a clear demarcation line between those who can decipher the code and those who can only be awed by it. The description of biology's task in terms of cracking the secret of life has involved a shift over what counts as "life," as well as also over who has the power to define, control, or even own this life. This particular rhetorical gesture has also had some material effects: it has instigated a new politics of life management. Molecular biologists working together with computer experts have positioned themselves as new guardians of truth or holders of the key to the secret of life, but they have also taken a substantial share in what Michel Foucault describes as biopolitics: the sovereign power over life and death, exerted over the bodies of the population. We already discussed in chapter 3 how in the first volume of *The History of Sexuality* Foucault traces back the origins of biopolitics to nineteenth-century Europe, to the preparation of natality and mortality tables, the development of sanitation, and the general management of "public health." This form of politics has arguably gained a new form and intensity in the biotechnological era, with the life and health of the populations becoming an object of a permanent attention of different sovereign forces: politicians, scientists, pharmaceutical companies, life coaches, the media.[24] The scientific process of revealing the structure of DNA thus constituted a significant episode in the political colonization of the private realm of the flesh, with all its ambivalent imperial undertones. This form of colonization entailed, first, the transformation of individual lives and bodies into a calculable biological entity described as a "population," and, second, the inclusion of this newly emergent "population" at the center of sovereign power—an inclusion that, in case of genomics, was nevertheless occluded by being presented only as a scientific discovery, not an act of control. The development of sociobiology,[25] a discipline involved in studying the behavior of social groups by using tenets from population biology and genetics, and its fine-tuning of the eugenic tradition, is one example of the role that molecular biology has played in this coming together of life and politics.

The biopolitics of "life in the molecule" has been accompanied by the arrival of another, globalized form of biopolitics: sovereign control and command over local or national communities has shifted to embrace the world population, since the early 1990s "unified" under the auspices of the Human Genome Project and the allegedly more inclusive Human Genome Diversity Project.[26] This explicit incorporation of biological life as an object of political and media attention has not been value-free or unproblematically beneficial. As discussed in chapter 3, in his engagement with the Foucauldian notion of biopolitics, the Italian philosopher Giorgio Agamben explores the vicissitudes of this "inclusive exclusion" of life by proposing the term "bare life," *la nuda vita*, which can also be translated as "life as such" or "life itself."[27] Bare life becomes for him a limit concept situated between natural life, the simple fact of living, and organized or already politicized life. This *zoē–bios* opposition implies another one—that between *oikos* (home, or the private realm) and polis (city, or the public sphere). As Agamben explains in *Homo Sacer*: "the inclusion of bare life in the political realm constitutes the original—if concealed—nucleus of sovereign power. *It can even be said that the production of a biopolitical body is the original activity of sovereign power*. . . . Placing biological life at the center of its calculations, the modern State therefore does nothing other than bring to light the secret tie uniting power and bare life, thereby reaffirming the bond . . . between modern power and the most immemorial of the *arcana imperii* [state secrets]."[28] The critical task today, we may suggest, is to expose this secret uniting power and bare life, a secret which remained hidden among the scientific revelations about the DNA structure. Scholars in science and technology studies, feminist studies of science, media and cultural studies, as well as the hard sciences have undertaken the interrogation of the relationship between politics and life in their work over the last few decades, but it is through Agamben's reading of the zone of indistinction between bare life and the political realm, where *zoē* is included in *bios* only through its exclusion, that this critical interrogation of life secrets as *arcana imperii* can be undertaken.[29]

We can see from the above how the genetic paradigm of life—where life is seen as reducible to the sum of the organism's genes—is established as a "common law" (which is also a "common sense law") and how it gains legislating powers over the lives and bodies of a population. However, its normative functioning remains a secret, or rather the genetic paradigm enters a zone of indistinction between being a description of what life is and a prescription of what it should be.[30] What holds life and law together, according to Agamben, is the relation of a ban through which "he who has been banned is not . . . simply set outside the law and made indifferent to it but rather abandoned by it, exposed and threatened on the threshold in which life and law, outside and inside, become indistinguishable."[31] In the context of life defined as a sequence of DNA packets, we can perhaps suggest that it is corporeality and bodily marked sexual difference, or rather humans as corporeal and sexuate beings, that constitute a ban to the genetic law (although I will later consider some other, "secondary" bans imposed and demanded by this law). Naturally, chromosomal difference is important to genetics—any high school biology student knows that each human cell contains twenty-three pairs of identical chromosomes in both men and women, with the twenty-third pair determining the difference between the sexes (in males, the twenty-third pair contains two unlike chromosomes, called X and Y, while in females both members of homologous pair no. 23 are X chromosome). Genetic difference is also the subject of the investigation of the Human Genome Diversity Project. As its chief proponents put it, "By an intense scrutiny of human diversity, we will make enormous leaps in our grasp of human origins, evolution, prehistory, and potential."[32] However, this is difference which is understood only as opposition and complementarity (between male–female) or as local variation (between different alleles of the same gene, affecting, for example, skin, hair, and eye color), not as an irreducible difference from itself, a crack within the identity of the self-same. Rosalyn Diprose argues that modern genetics ultimately attempts to efface differences and claims that "genetic theory is itself a genetic operation—it is involved in the production of difference in terms of sameness."[33] Even though, through complexity theory applied to biology, structural and functional differences between organisms are foregrounded, "all these differences are sidestepped when we consider the nucleic acid sequences [i.e., RNA and DNA] from whence all creatures derive."[34]

The recoding of life as a bioinformatic secret has not just resulted in the foreclosure of difference through its disembodiment or desexualization; it has also involved the wresting of life from women.[35] Cracking the secret of life has been presented as a victory of male productivity over female procreativity,[36] as the overcoming of messy "stuff," of flesh and blood, and the illumination and purification of Nature. This move away from "life itself" has led, according to Keller, to a closure of a gap between lifeless and life-destroying forms: nucleus cells and nuclear energy, babies and bombs.[37]

This does not mean, however, that the secret of life is out in the open. The exclusion of the dirt of the matter from the scientific conceptualizations of life has only shifted the boundary of the licit and the illicit, the speakable and the unspeakable. The originary relation of a ban that excludes difference as foundational to life by recoding it as mere complementarity, and that relegates life's meatiness to the epiphenomenon of cellular processes which can nevertheless be ignored through a change of scale, has also established a series of what we may call secondary bans, whose working is premised on secrecy. For Agamben, a ban is an act of naming someone or something through cursing them (cf. the etymology of the word *bannan*—to summon—in the Old English) and placing them outside the licit, but it works through concealing its operation of power. In other words, the fact that the process of exclusion is also a simultaneous inclusion of the banned in the relation of exception needs to remain secret. Significantly, the secret itself works according to a similar principle. Originating from the Latin word *secernere*, which means to separate or shift apart, it is an act of drawing boundaries, of excluding something to the margins while also activating the unstable tectonics of the margins, where the provisional and always contingent centrality of the included is only maintained by depending on its "constitutive outside," on what it is not. It is also, contra Agamben, a process of establishing a relation with alterity and, thus, setting up an ontology of foundational difference that poses a challenge to Agamben's philosophy of immanence.[38]

The Secret Is Out . . . Long Live the Secret!

There is no room for such philosophical vacillations, though, in the discourse of one of the most prominent "crackers" of life secrets, James Watson. Watson claims that molecular biology has shown us clearly and unambiguously that "life itself was not complex, as had been thought, but simple—indeed simple beyond our wildest dreams. The only secret of nature was that there were no secrets, and now that secret was out."[39] While the secret of molecular biology is thus established here as being out there in the open, as a plaything of the scientist who celebrates its cracking as soon as he declares its secrecy, it conceals but also generates many other secrets. And it is these other secrets, beyond the "exquisitely organized physics and chemistry,"[40] that are more important from the point of view of engaged cultural criticism. How can we begin to disentangle this underlying web of secrecy that manages to conceal some of the vital questions about "life" precisely via the scientific rhetoric of revelation and transparency?

In addition to the issues of corporeality and sexuality discussed above, we should mention the secrecy issues brought up by the development of genomics and genetic testing in the aftermath of the cracking of the DNA code. Indeed, over the last few years we have witnessed the rise of the material–spiritual bio-industry that promises to help us reveal our genetic secrets, allegedly "for our own good." And yet it is worth asking whether we would want it to be revealed, for example, that we are carrying a gene for a disease for which there is no cure.[41] How would we handle the secret of being a carrier of a "breast cancer gene"—would we opt for preventive double mastectomy or carry on as usual in the hope of beating the odds? In "The Secret of Life" issue of *Time* Nancy Gibbs raises a further question regarding genetic testing in the context of privacy and its links with individual and corporate responsibility: "Can it really be kept a secret from your boss or your insurance company or your future spouse that you carry a gene that predicts you will develop Alzheimer's by age 45?"[42] A thorough consideration of all these questions would, of course, require a detailed study of particular diseases and the possibilities of their prevention and cure (testing couples preprocreation for the possibility of Tay–Sachs disease would bring up different issues than testing fetuses for Huntington's disease or adults for Alzheimer's); it would call for the interrogation of the political ideologies that support individuals' right to "privacy" while also positioning them as bearers of corporate responsibility; it would demand questioning what kind of life is considered worth living and what kind of values underlie this concept of life; and, last but not least, it would require us to take a closer look at the financial interests and ideological positions of the gentech companies involved in the business of cracking the secrets of life. All these issues would open up into a whole lot of new questions and singular contexts, which could not possibly be dealt with within the space of this chapter. I am only raising them here in order to indicate that such a difficult, painstaking interrogation is necessary in a discussion of the politico-ethical issues concerning bodies and lives. Any attempt to develop a clear policy of bioethical regulation prior to the consideration of the above issues (a consideration which will inevitably be ongoing)—no matter if conducted in the name of moral responsibility, defense of human rights, or political pragmatism—will only thicken the veil of secrecy that envelops "life."

The big secret of interests and investments, both affective and financial ones, in the biotech–gentech industry deserves separate consideration. There already exists a substantial body of work critiquing the corporatization of the biotech industry and raising objections to its politics and ethics: one can think here of Vandana Shiva's *Biopiracy: The Plunder of Nature and Knowledge*, Jeremy Rifkin's *The Biotech Century*, Donna Haraway's *Modest_Witness*, or Sarah Franklin et al.'s *Global Nature, Global Culture*. However, the affective investments raise as many questions. The feminist critic Mary Jacobus argues that the quest for cracking the secret of the DNA code could be seen as driven by erotic energy, with life positioned as an object of male desire—less for itself but rather because it was desired by another scientist. Both

Jacobus and Keller analyze the competition between the scientific teams on the way to conquering the deepest entrails of nature and tearing out her secrets in terms of a stereotypically gendered sexual conquest, with the metaphorical elision of woman from the story of life accompanied by the actual erasure of Watson and Crick's Cambridge colleague Rosalind Franklin from the official story of their scientific triumph.[43] We could also seek parallels between this earlier adrenaline-driven quest for life and the Human Genome competition between the U.S.-government-backed team and J. Craig Venter's corporation Celera Genomics over who will produce (and, subsequently, own and manage) the "map of humankind" first.

As Deleuze and Guattari recognize in another context, "the secret always has to do with love, and sexuality."[44] The cracking of the secret of life, its redescription in terms of mother molecules, originally involved a de-Oedipalization of family ties, a liberation of life from traditional kinship structures, and the establishment (via programs such as the Human Genome Diversity Project) of other forms of international and interracial community and belonging. Nevertheless, the potentially radical consequences this severing of family ties might have had for sexual and gender politics was quickly scuppered by the immediate transference of the love interest to the corporate structures. Rather than build "inoperative communities" based on new forms of nonbiological, nonessentialist kinship, the "hackers" of the secret of life quickly passed on this secret to the biotech industry, whose own desires and libidinal investments are primarily driven by the flow of capital. Thus, the question of investments, interests, and corporate obligations of biotech companies entails a call to probe to what extent libidinal energy drives the business of the decoding of life. Also, from a critical feminist perspective, it is worth mentioning that the Oedipal relation of paternity seems to be reestablished once the disembodied and disentangled secret of life enters the corporate world, once it is encoded again—this time as a node in the economy of biopolitical production.

There is another layer in this web of secrecy, where the ideality of the secret is enacted through its unnameability. This is the secret of a life for which we either yet do not have a name (new "soft" cyborgs—

children of OncoMouse and Dolly the sheep, creatures of unknown provenience, human–machines, artificial life forms) or whose name we dare not speak (asylum seekers, victims of torture, "local" civilians dying in the Iraq war as part of the "collateral damage"). Situated outside the vital network of social and political recognition, already in an uncanny proximity to death that life in the polis sanitizes and denies, these "precarious lives" do not have a place within the traditional political discourses.[45] They even remain outside the category of a "problem" that needs to be resolved and managed. These are immaterial lives, true ghosts haunting the democratic polis with their absent presence, with their—to borrow Agamben's Auschwitz-inflected term which has gained a new resonance in the context of current Islamophobia—*Muselmann* exceptionality.[46]

It might seem that from a political point of view, it is more important to reveal the secret of the second group, that is, to claim political legitimacy for those whose lives do not seem to occupy a prominent position in our economy of biocalculation—asylum seekers, victims of torture, and so forth—rather than explore the livability of cyborgs or alife creatures. Also, it might equally seem as if I have just turned from a narrower, more specific definition of life as "code" to its broader understanding in terms of livability, the possibility of biological and political existence. However, it is worth pointing out that the metaphysical idea of life as transcendence shapes the mechanistic concept of life as information, or code, in the life sciences—even if, in case of the latter, it is placed in the position of negativity, both evoked and denied through an act of reflective mirroring.[47] And thus all these issues, bodies, and levels of life are not unconnected. The consideration of the "unlivable lives" of those banned or forgotten by democratic states is, of course, vital to the ethical project outlined here, but it cannot be undertaken without a more thorough interrogation of what counts as (human and nonhuman) life, and, in particular, of the economy of calculation that informs both the dominant forms of capital-invested politics and market-driven bioethics (the latter usually presented within the framework of utilitarianism). Neither can it be attempted without looking at the individualistic premises of dominant politico-ethical models in the capitalist world, where the relationality of living beings is overlooked for the sake of the analysis of monadic entities, and at the expense of the forgetting of flesh, of sex, of sexual difference. The very process of the rhetorical conceptualization of life in science is thus of interest to me here because it both underpins and drives the current democratic biopolitics of life management. It is precisely the perception of life in terms of information flows and codes that is significant to the rearticulation of politics in biological terms, to it becoming a biopolitics. This process also involves the dematerialization of life, which in turn facilitates the bracketing off of those who are not deemed worthy of membership in the category of the living. While licit biopolitical citizens have their identity and political belonging reconfirmed through a series of prosthetic data extensions—biometric data cards and passports, magnetic strips on their credit and store cards—those whose names we dare not speak function only as white noise in the computational system of life management, as a disturbance to be watched out for in order for it to be eliminated (e.g., in the form of a digital image rendered by a surveillance camera installed in the Channel Tunnel between England and France).

By turning to the secret of unnamable and unspeakable lives I have not therefore shifted my critical attention from the micro- to the macrolevel, or replaced the definition of life as information with a more abstract, philosophical, or even fuzzy concept. Indeed, a neat separation between these different concepts of life is not possible, as the very articulation of life as code activates the working of exclusionary mechanisms at the (bio)political level of the state and its institutions—such as welfare agencies, asylum and immigration centers, counterterrorism cells, DNA testing laboratories, and so forth—that rely on information and data in order to process, survey, and control the state's subjects. One of the secrets behind the promotion of the "secret of life" trope, where the secret's cracking lies within scientists' power, is precisely the elision of the more messy, material as well as political, aspects of life that cannot be neatly rendered as a sequence of DNA base pairs or codons.

If the mathematical and information sciences have provided one of the dominant paradigms for thinking about human life today, the recoding of life as a sequence of data that holds no secrets, a secret that has already been cracked, has facilitated the multilevel politics of life management and control—from the cell to the organism as a whole. The question of cyborgs, human–machine, material–virtual life forms that trouble our traditional understanding of what it means to be alive, therefore needs to be considered together with the question of unnamable, excluded lives whose function is to conserve the legitimate boundaries of the democratic polis. The work conducted in science and technology studies, feminist studies of science, and media and cultural studies in particular can prove useful in tracing the parallels between the working of biopower on the minds and bodies of the inhabitants of both "virtual" and "real" worlds. But it can also allow us to query the demarcation of virtuality, as well as ask further questions about the invisibility and unspeakability of certain forms of life that do not fall into the clearly delineated binary. Judith Butler warns us that "There is always a risk of anthropocentrism here if one assumes that the distinctively human life is valuable . . . or is the only way to think the problem of value. But perhaps to counter that tendency it is necessary to ask both the question of life and the question of the human, and not to let them fully collapse into one another."[48]

A Bioethics of Corporeal Obligation

If the political today is caught between a ceaseless constitution of human and nonhuman, and human and animal,[49] it seems to me that an attempt to crack the secret of life has involved not only reconceptualizing living beings as pockets of DNA but also endowing different "pockets" with differential values. We could perhaps repeat here, after Agamben, that within this zone of (DNA) indetermination, the human has become "the place of a ceaselessly updated decision in which the caesurae and their rearticulation are always dislocated and displaced anew."[50] I would like to suggest we adopt this formulation as a new line for bioethical thinking, which will function as a response to the current biopolitics and its underlying genetic law. The indeterminacy of the human, coupled with the need to make a decision about the status of human, nonhuman, and inhuman life constantly anew, can be seen as a positive obligation driving this new ethics of life (an obligation that perhaps goes beyond some of the negativity of Agamben's own diagnosis of "the human condition" in *Homo Sacer* and *The Open*). This obligation places a demand on us to respond to this indeterminacy of living beings and not to foreclose it in advance—be it by means of a humanistic belief ("human life is sacred") or scientific calculation ("human is a mere collection of data"). And thus the original structuration of life in terms of the secret in molecular biology, waiting for "experts" to reveal it while concealing its own foundation or its "constitutive outside," needs to be taken issue with in this context precisely because of what it excludes. This is particularly important because, as Sarah Franklin observes, it is the biogenetic definition of life that informs many moral debates today.[51]

An alternative bioethics thus requires a new entry point, outside the traditionally delineated discipline of moral philosophy. I would like to suggest that cultural studies in particular—with its ethical imperative to respond to and take responsibility for incalculable difference, and its critique of institutions, capital, and power from Marxist and post-Marxist perspectives—may provide a valuable perspective for conducting this sort of investigation. Cultural studies can help us intervene into the network of discourses of secrecy and transparency surrounding issues of life with something we might term "an ethics of life's countermanagement" that will also consider irrational investments, incalculable spending, and the contingency of life that nevertheless refuses to remain tied to the biocapitalist machinery. The theoretical and political frameworks developed under the aegis of cultural studies seem particularly useful in addressing the question of life's secrecy and the legitimate and illegitimate participation in both its definitions and secrets because, as Clare Birchall has put it, "As an originally oppositional discourse that still has a precarious status within the university (despite an undeniable institutionalisation in some locations)—cultural studies has a greater capacity for opening itself up to questions of legitimacy than others. Cultural studies is well placed to 'expose' rather than 'keep' the secret of undecidable legitimacy: a secret that conditions any knowledge statement, and anything that we could recognise as cultural studies."[52]

In its interrogation of the legitimacy of the genetic law, of the reductionist thinking which equates life with DNA information, this new bioethics will be "experimentally concerned with what 'bio' means in relation to 'ethics,' with the ways in which ethics always involves a 'bio' component."[53] Considering the very definition and conditions of life that are both its driving force and regulated object, this bioethics will also allow us to cast light on the link between violence and the law, and between power and bare life, as the secret of biopolitics that organizes life in Western (and increasingly other, or "newer") democracies today. For Agamben, in the modern state the secret is already out in the open—and yet it seems paradoxically concealed precisely thanks to the ubiquitous, all-pervasive visibility of biotechnological machines that constitute, manage, and exclude different forms of "life itself." The secret of life thus seems analogous to what Birchall has described as the Derridean secret: "that which remains outside the phenomenal event as it happens but which nevertheless conditions that event [i.e., "life itself"]. The irreducible, nonpresent secret (or in fact 'nonpresence') in this sense structures presence. I can name this secret 'undecidable legitimacy' or something like that, but this is really only akin to saying the secret is that no-one knows the secret. The secret remains irreducible even while we try to reveal it, keeping the future open."[54] Another way of articulating this paradox of the secret of life is by saying that life as revealed to us in the discourse of the life sciences depends on secrecy or that it is in fact the product of a secret. We may go so far as to suggest that it would indeed be impossible to think life without or outside secrecy, because the secret is the very condition of its conceptualization. As Richard Doyle puts it, "the great unsaid of the life sciences, of a molecular biology that sought and found 'the secret of life,' is the fact that life has ceased to exist. Or, rather, that it never did exist, that the life sciences were founded on an embarrassing but productive ambiguity, the opaque positivity called 'life.'"[55] The "secret of life" trope used by both science and public discourses about science conceals the fact that the privileging of mechanism over vitalism in its definition of life as code is itself a mechanistic division. It is a structuration within language which temporarily keeps apart such terms as materiality and virtuality, matter and spirit (terms which are mutually implicated in each other as their conditions of possibility). Any attempt to reduce or reveal the secret of life—even, or especially, if conducted in the name of science and with the application of all of science's legitimating apparatuses and protocols—will only end up displacing the secret, rearticulating it, while also concealing science's own situation in the networks of power and the law (the latter having illegitimacy and originary violence as its condition of possibility).[56]

If "cracking" reveals itself as a somewhat futile approach to the secret of life, we should therefore rather focus our attention on the more doable and practical problems that this secrecy commands. It is in the thinking of an alternative bioethics, as explained above, that I locate this pragmatic task of addressing the secret of life otherwise. Responsible bioethical thinking today needs to involve looking at the secrets of the state, of biotech corporations and of scientific knowledge protocols that are already out in the open, or at least in the process of being drawn, but it also concerns the consideration of what constitutes a livable life and who has the power to determine livability. The recognition of the concepts, terms, and norms that govern life is important because, as Butler explains in *Undoing Gender*: "I may also feel that the terms by which I am recognized make life unlivable. This is the juncture from which critique emerges, where critique is understood as an interrogation of the terms by which life is constrained in order to open up the possibility of different modes of living; in other words, not to celebrate difference as such but to establish more inclusive conditions for sheltering and maintaining life that resists models of assimilation."[57]

The main task of bioethics today is thus not only to "concentrate on ethical challenges due to quite striking biomedical advances, such as those in genetics," as Stephen Holland's textbook of bioethics postulates,[58] but also to imagine different ways of naming, and thus recognizing, the new forms of life, without taking premature recourse to the rhetoric of assimilation or exception—via the tropological strategies of moral panics, monstrosity, or alienness. This approach requires the acknowledgement of yet another, perhaps most significant, secret of life, the secret of the alterity of the other (lives), of the fact that the other (life) does not yield itself to thematization, that there is always something that escapes my conceptual grasp. This recognition is the first condition of a new bioethics, an ethics of life that arises as a response to, and responsibility for, the secret that reveals itself before me and that imprints itself on my skin.

Speaking against the totalizing view of the world in which the self is the source of all meaning and the starting point of both morality and politics, while human life can be synthesized as a sum total of all events and experiences in which consciousness leaves nothing outside of itself,[59] Emmanuel Levinas defends the secret of life in the following terms: "The real must not only be determined in its historical objectivity, but also from interior intentions, from the secrecy that interrupts the continuity of historical time. Only on the basis of this secrecy is the pluralism of society possible. It attests this secrecy."[60] Secrecy here entails the abandonment of a desire to render a complete narrative of the other's life, of her past, present, and future. It is thus also a condition of both ethics and responsible politics. Ethics does not consist here of a code of morals—it will not tell us in a prescriptive and formalized way how to respond to challenges brought up by genetic testing, xenotransplantation, or cloning, as that would reduce ethical thinking to a reflection on totality and would thus prove ultimately unethical. Instead, ethics is seen as a primary demand or obligation posed by the infinitely other, whose otherness transcends the Leibnizian monadism or genetic complementarity and goes beyond difference within a genus. Significantly, for Levinas, "it is not as the individuals of a genus that men are together. One has always known this in speaking of the secrecy of subjectivity. . . ."[61]

One can hear echoes of Levinasian ethics as well as Hegel's notion of the self as *ek-stasis* in Butler's injunction, "We're undone by each other. And if we're not, we're missing something."[62] If we follow this injunction, we can interpret secrecy—which, as we know by now, is never absolute—as an ethical condition of being with others. Secrecy does not amount here to embracing passivity, ignorance, or lack of interest in relation to the other, nor does it mean egoistic self-enclosure against the other's reaching out to me. However, it does stand for the suspension of the possibility of totalization with any relationship to the other forms of life—human and nonhuman ones, those legitimately recognized as livable and those excluded from the conditions of livability. This suspension is only one aspect of my relation to other lives and bodies. Another one involves the need to respond, always in a singular way, to the demand that other lives and bodies place upon me, to the vulnerability and corporeal dependency that my being in the world always already implies. This form of bioethics can provide a political grounding for a new biotech era, in which the ontology of human and nonhuman bodies and lives is being redrawn. It can also allow us to think of some new forms of sociality, of being with radically others, where their alterity is not immediately absorbed into the biopolitical machinery of technocapitalism, with human and nonhuman lives all assigned a biovalue that determines their visibility and legitimacy. Even though Levinas's ethical thinking focuses on the human other, it is from the idea of infinity as God, and the other human as if he or she were God, that openness to other forms of life for which we do not yet have a name can be derived. This openness can create possibilities for ethical politics, "the pluralism of society" Levinas talks about, which would not be rooted in self-focused and interest-driven liberalism but rather in what Philippe Nemo describes in his conversation with Levinas as "the essential secrecy of lives."[63] This notion of secrecy, based on the conviction that the social results primarily from the limitation of the principle that "men are *for* one another," is opposed to the concept of the "secret of life" discussed at the beginning of this chapter, which posits society as "the result of a limitation of the principle that men are predators of one another," subject to the ultimately cognizable laws of nature.

This is by no means to defend something like an intrinsic dignity of the human or the sanctity of life. Secrecy stands here rather for a pragmatic recognition of limitations of the self, any self, and of its both ontological and epistemological dependence on other bodies, lives, and nonmaterial entities. This is why, even if I have reservations about its anthropomorphism, I nevertheless want to suggest that Levinas's theory of sociality which is not predicated on the identity of the self-same would be ethically *better* than the genetic paradigm, which posits life as a soluble and systematic task while seeing human society as a genus, a Family of Man united under the auspices of the Human Genome Diversity Project. Even if Watson assures us in the last chapter of *DNA: The Secret of Life* that the essence of humanity is love and that its inscription in our DNA "has permitted our survival and success on the planet,"[64] his hope that someday "particular genes . . . could be enhanced by our science, to defeat petty hatreds and violence"[65] risks obliterating the boundary between a love parade and a prison camp. In order to avoid entering such a zone of indistinction between the protection of innocent lives and the regulation of disorderly ones which would be run by the biopolitical machinery of life management, it may be worth taking account of Levinas's story of the secret.

6

Green Bunnies and Speaking Ears: The Ethics of Bioart

The *Matter* of Art Today

Contemporary artists' engagement with new art materials—human and animal tissue (Oron Catts and Ian Zurr of SymbioticA),[1] human fat (Stelarc and Nina Sellars),[2] genetic strands (Critical Art Ensemble; CAE)—charts new territories in the landscape of art history and problematizes even further the already contested distinctions between subject and object, art and life. The so called "bioart," which stands for art utilizing biomaterial such as tissue, blood, or genes as its medium, evokes a lot of controversy, both within and without the art community. Typical responses to bioart reflect a wider public anxiety regarding the current transformation of human and nonhuman life and its mediation by technology. This anxiety can be manifested in the not infrequent public evaluation of art practices that actively engage with biomaterial in terms of a scandal, a violation of the dignity of the human and of life as we know them. Such reactions in turn serve as a breeding ground for a moral panic about our "Frankenstein world."

However, the controversy aspect of these kinds of art practices is only one element of the debate on the ethics of bioart that concerns me in this chapter. I also want to argue here that, even if we concede that artistic experimentation with new "soft" technologies can be seen as thought-provoking, the actual theoretical discourse that accompanies such experimentation is often either excessively didactic and moralizing—claiming that art should work in the service of politics and ethics—or too deterministic, too prone to techno-hype and uncritically fascinated with the technological process itself. This chapter thus asks how we can counteract the moralist responses to, and justifications of, bioart on the part of both artists and their audiences. It also offers yet another attempt to think life and bioethics otherwise—this time with the help of particular art practices—in a way that goes beyond the predefined discourses of moral anxiety, artistic liberation, or technological determinism.

The artistic interrogation of the relationship between the body and technology is well-documented: we can list here, for example, Orlan's radical plastic surgery series, *The Reincarnation of St Orlan* (1990–93), Stelarc's robotic installations which include *The Third Hand* (1976–81) and his Internet actuated and uploaded performance, *Ping Body* (1996), as well as Eduardo Kac's telepresence work such as *Rara Avis* (1996). In all of these projects the symbolic and material boundaries of the human have been opened to technology in a series of gestures that some consider hospitable and others offensive and sacrilegious.[3] These "humachinic" encounters allow for an enactment of a networked relationality between bodies, organs, and objects and, consequently, for a development of new models of codependence and co-evolution that challenge the monadic, human-centered understanding of the world.

Art Meets Science: The "Public Understanding" Agenda

The last decade or so has seen the arrival of newer, smaller, and softer technologies on the art scene. Partly reflecting the broader public debates on cloning, gene therapy, GM foods, and bioterrorism, and partly responding to the shift toward biotechnology in scientific committees' and funding bodies' agendas (as evidenced by high-visibility programs such as the Human Genome Project), artists have taken on board the discourses and practices of the scientific lab. With the use of biomaterials—tissue, blood, genes—in artwork, the old adage "Art is Life" has gained a new meaning. Artists' engagement with biotechnology has also inevitably become a pedagogic process. It requires both theoretical and practical self-education in the recent developments in biosciences on the part of artists themselves and the subsequent transmission of this knowledge to the public: sometimes in the form of instructions as to what materials are being used and what they actually "do," often—as in the case of CAE—in terms of a wider discussion of the current scientific practices and their ideological and social implications.[4] Educating the public about science becomes for many artists a way of demystifying not only the technology itself but also the intricate financial interests and investments that underlie the biotech industry. Natalie Jeremijenko and Eugene Thacker's innovative publication, *Creative Biotechnology: A User's Manual*, develops the idea of the "biotech hobbyist," encouraging lay audiences to engage with biotechnology in a playful, amateurish way. (The book even comes complete with instructions on where to purchase mice for experiments and on how to grow your own skin!) In the context of the moral panics over terrorism and biohazard post-9/11, this is, of course, a risky strategy. However, it is precisely the desire to "decouple biological and military concerns"[5] and thus, we may say, inject some life back into Life through the defense of "the right to tinker,"[6] that drives the biotech hobbyists' agenda. Life here does not stand for the precious substance that deserves particular respect due to its metaphysical connotations. It is rather life as ubiquitous raw matter which is already part of us but which also invites us to get our hands and minds dirty, to play with it and think with it, that is the subject and object of this particular "creative biotechnology" project.

The pedagogic aspect of many bioart projects is not without problems, though. This agenda for promoting the "public understanding of biotechnology," which is so explicitly espoused in many bioart projects, can make artists adopt a somewhat servile role toward biotech and bioscience, with art becoming a mere handmaiden to science. Despite the critical tones of many artists' pronouncements, projects of this kind, often legitimized by "art–science collaboration programs," whereby a bioartist is allowed into a scientific lab for a short- or long-term residency, may actually end up reaffirming the authority of biotech as well as the disciplinary hierarchy. Eduardo Kac's declaration that one of the principal ideas behind his GFP Bunny project, where GFP stands for green fluorescent protein, was to facilitate "ongoing dialogue between professionals of several disciplines (art, science, philosophy, law, communications, literature, social sciences) and the public on cultural and ethical implications of genetic engineering"[7] raises precisely such doubts. Kac's GFP Bunny was a transgenic albino rabbit, created in a French science lab with the help of EGFP, an enhanced version of the original wild-type green fluorescent gene found in the jellyfish *Aequorea victoria*, as a result of which the animal was supposed to glow when illuminated with a particular type of ultraviolet light. In an article on this project, Kac reassures the audience that no genetic mutation is effected through the transgene integration of EGFP into the host genome and that the inserted substance is entirely harmless to the rabbit. He is also keen to point out that this work "does not propose any new form of genetic experimentation, which is the same as saying: the technologies of microinjection and green fluorescent protein are established well-known tools in the field of molecular biology. Green fluorescent protein has already been successfully expressed in many host organisms, including mammals."[8]

The recontextualization of the established scientific procedure, conducted with the aim of opening up a debate about DNA, biodiversity, purity, hybridity, and breeding, evidently inscribe Kac's GFP Bunny project in the pedagogic framework. The artist here attempts to challenge the accusations that he is merely being a scandalist, and takes on instead the role of an educator, intent on instigating a public discussion over the politics and ethics of biotechnology. However, we could ask here, together with Krzysztof Ziarek, "whether art is actually needed in order to generate the kind of discussion, no doubt crucial and imperative, that has been going on around Kac's work, or whether those questions do not in fact arise from the very premises, objectives, and capabilities of genetic technology."[9] While acknowledging that Kac's work has accelerated this debate and provided new insights into our understanding of genetics, Ziarek goes so far as to actually question whether this is enough to call it a work of art. He adds: "In the context of this thinning boundary [between technology and art], it seems legitimate to ask whether and to what extent transgenic art is complicit with the manipulative flows of power or whether . . . it exposes, complicates, or perhaps even contests them."[10]

I want to suggest that such potential complicity lies, paradoxically, in the inherent moralism of many bioartists' convictions and articulations. By "moralism" here I mean a clearly outlined and firmly asserted political position whose foundations and conceptual or geographical situatedness (usually "on the outside" of the institution, the hierarchy, traditional politics, etc.) do not seem to require any justification because they are presented as self-evident. This logic and discourse is, of course, not the prerogative of bioartists: it can be found in many different political or artistic positions: left- or right-wing political activism, subcultures, or tactical media. What all these positions nevertheless seem to have in common is not just a clear identification of the enemy—in the case of bioartists, the biotech industry—but also the obfuscation of the conditions of emergence of their own position, or of what I called in my earlier work, with a nod to Ernesto Laclau, "ethical investments."[11] Such investments are predominantly affective and unconscious, but they are also necessary for outlining any project or program, be it political or artistic. In other words, to engage in any artistic practice or to undertake any form of political action, we need to be guided by some prior belief ("an ethical investment"), which belongs to the order of "universals." This ethical investment usually takes the form of an unacknowledged attachment to a certain idea of truth (artist as social conscience, as revolutionary, as emancipator) and to identity conceptualized in terms of injury (artist as part of the wronged social group which sets itself against "society" and "capitalism"—the socially deprived, the creatively hampered, etc.). And yet the partly unconscious and unacknowledged nature of this investment for which no account is given can lead artists and activists to adopt moralism, "a kind of posture or pose taken up in the ruins of morality by its faithful adherents."[12] This is to say that, when morality is seen as too oppressive, too compromised or too outdated, but its proponents are not prepared to analyze the philosophical and ideological conditions of this situation, it tends to "go underground" and turn into its ugly twin: moralism. Wendy Brown argues that moralism ends up in an antagonistic position toward a richly conflictual political or intellectual life. Ultimately, it takes the form of antipathy toward politics as a domain of open contestation for power and hegemony, even if the moralist's discourse is full of appeals to "politics," "intervention," and "justice." Moralism in art and politics can be pernicious as it replaces the passion of quasi-religious conviction, which can nevertheless drive a liberatory movement, with paranoia, mania, and, ultimately, political stasis.[13] In the last instance, a moralism that remains unaware of its own ethical, affective investments turns out to be both anti-political and anti-ethical.

This is why I am somewhat suspicious, for example, about Natalie Jeremijenko's pronouncement that bioartists have "no business in biotech" and that they are "unique in the biotech realm, which is largely directed by corporate interests."[14] While the exploration of the power structures of the biotech industry and its influence on the overall medicalization of populations by both artists and activists can certainly be seen as politically significant, Jeremijenko seems to overlook artists' ethical investments in the cultural capital in which they trade, that is, in the modernist idea, perhaps even fantasy, of the artist as an outsider to the dominant regimes, as well as the "alternative" or "tactical" discursive and political power wielded by their own narratives about biotechnology. Jeremijenko goes on to claim that "Impartiality is a precious commodity in the biotech world and one possessed primarily by a small group of artists."[15] Again, this sort of argument is reminiscent of broader debates among "the left" or "the progressives" regarding the radical opposition to capitalism as a delimitable entity from which the political agent can isolate itself, and the accusations of "cooptation" directed at the proponents of more networked and more embedded political models.

Returning to Kac's GFP Bunny, if the novelty of this particular art project does indeed lie in the political and ethical debate it generates more than in its aesthetic dimension, it seems to me that we are posed with a risk of this very project being scuppered all too early by precisely such a kind of moralism—on the part of the artist who *knows in advance* that this is an important scientific issue that the public needs to be exposed to and on the part of "the media" (i.e., the project's producers, audiences, and the broadly conceptualized "art world") who will bring to the table their ready-made values regarding nature, culture, and technology. This is not to suggest a return to some unreconstructed aestheticism in which we should just wonder at the beauty of the green rabbit (or the lack of it); it is only to suggest that we should perhaps look at what Ziarek calls art's *forcework*, the concept that springs from his Heidegger-, Foucault-, and Deleuze-inspired understanding of art "as a force field, where forces drawn from historical reality and social reality come to be formed into an alternative relationality."[16] What Kac's GFP Bunny fails to achieve for Ziarek, then, is to enact this transformative event: it does not do anything that sciences themselves would not have achieved beforehand.[17]

Interestingly, in the Conclusion to his book, *The Global Genome*, Thacker seems to concur with Ziarek's conclusion regarding the uncertain ontological and epistemological status of bioart when he states: "I find it worthwhile to refuse this convenient tag [of bioart] for a number of reasons. It not only marginalizes (or niche markets) art, effectively separating it from the practices of technoscience, but the notion of a 'bioart' also positions art practice as reactionary and, at best, reflective of the technosciences. Though the term *bioart* may indeed refer to artists and artworks circulated primarily within the gallery system, we might do better and ask how cultural research dealing with biotechnology can take seriously its interdisciplinary nature."[18] While Ziarek applies rather formal modernist criteria in his evaluation of Kac's work, allowing or even clamoring for art's special functioning "otherwise than through dominant articulations of power,"[19] Thacker willingly acknowledges bioartists' (if we can still call them this) complex interdisciplinary engagement with the world of technoscience and with the power structures and flows associated with this world. Aware of the limitations of the activist, "tactical media" discourse that surrounds many bioart practices, where artists creatively work through their own space of exclusion from the dominant institutional and political frameworks, Thacker is prepared to go beyond this exclusion–inclusion, oppression–incorporation rhetoric and concentrate instead on bioart's "*tactical effects* that often leave only traces of their successes to be discovered later by the ecologists of media" (quoting Alex Galloway).[20] We can perhaps see a similarity between Ziarek's "forcework" and Thacker and Galloway's "tactical effects"—the difference that emerges here is between these different theorists' critical evaluations of bioart. The question of bioart's effects, or of what Ziarek describes as "a becoming, in which forces unfold through each other in a continuously reactivated field of reciprocal shaping, because in this type of articulation all forces are both affected and affecting,"[21] leads us to another important question: how do we actually assess that its forcework, or tactical effects, have indeed taken place? Also, is this assessment the task of an art critic, a media ecologist, a user–participant, a policy maker, a bioethicst?

Beyond Moralism: Bioart and Bioethics

The proposition I would like to advance in the further part of this chapter is that the tactical effects of many bioart projects lie in their ability to shift the discourse on, and of, (bio)ethics,[22] provided ethics is understood, as it has been throughout the book, in the Levinasian sense as *primordial* and as exerting a demand on the political and the social. In other words, the forcework of bioart projects consists in their ability to outline, even if not always consciously or explicitly, a different discourse on life in its politico-ethical contexts, as well as its economic and noneconomic valuations. The purpose of this alternative is to challenge many of the dogmatically grounded, moralist positions on bioethics and life, in which the power of certain political convictions and economic interests is obscured by the rhetoric of universal values or by particularist assumptions. I want to argue that artistic experiments with biotechnology and genomics not only contribute to the ongoing political debate on these values and assumptions but can also be seen as challenging the dominant biopolitical regime. In other words, I suggest that even if their rhetoric may be rather moralist, the forcework of many bioartists' projects and performances can indeed be ethical—a conclusion that will allow me to propose a more sympathetic reading of Kac's art in the further part of this chapter.

The ambivalent relationship between moralist pronouncements and ethical forcework can also be identified in the work of CAE. CAE position themselves as both artists and activists, whose task is to "resist . . . authoritarian structures" of the dominant biopolitical regime and envisage a new techno-apparatus which will counter the "obsessively rational" military and corporate drives of "pancapitalism." Their pronouncements often take on a prophetic tone, which can be heard in the following passage: "If big science can ignore nuclear holocaust and species annihilation, it seems very safe to assume that concerns about eugenics or any of the other possible flesh catastrophes are not going to be very meaningful in its deliberations about flesh machine policy and practice. Without question, it is in the interest of pancapitalism to rationalize the flesh, and consequently it is in the financial interest of big science to see that this desire manifests itself in the world."[23]

Even if there is something rather problematic about too quick an assumption CAE make that human matter should never be sold[24] and that "pancapitalism" is an identifiable network of (evil) forces against and outside which artists can take a clear and principled stand, the playful aspect of their projects—*Flesh Machine*, *Cult of the New Eve*, *Free Range Grain*—highlights the deeply problematic role of the biotechnological industry in shaping the current political and social consensus about the "value" of life. *Flesh Machine* (1997–98) consisted of a series of performances intended to highlight various contentious aspects of new reproductive technologies. Participants were invited to take donor screening tests and gather information about reproductive technology, as well as consider the sociopolitical aspects of such procedures. The *Cult of New Eve* project, performed in international art galleries in 2000, attempted to examine "the appropriation of Christian promissory rhetoric by industry and scientific specialists in order to persuade the public of the utopian nature of new technology."[25] The artists took on the scientific discourse around biotechnology and playfully repositioned it as a form of a cult, drawing on biblical and mythical associations between femininity, nature, and life in order to raise questions about the authority and legitimacy of biotech today. *Free Range Grain*, a collaboration between CAE and two other artists, Beatriz da Costa and Shyh-shiun Shyu, was a series of installations for European audiences focused on questioning the permeability of state and trade boundaries in the age of globalization. The artists' role was to provide the public with adequate information—via stands, posters, and public food testing—about food technologies and the science and politics behind them. Their emphasis was on the global flow of GM crops such as corn and soy.[26]

CAE argue that in the times of "the flesh machine"—a Deleuze-flavored term they use to describe the current sociopolitical moment, in which previous military and visual elements of control have joined forces with medicine and bioscience—"ethics has no real place other than its ideological role as spectacle."[27] The force of CAE's writings, freely available on their Web site in the e-book format and presented in art galleries and other public spaces, raises important questions about science's "maniacal sense of entitlement," while also highlighting the fact that resident ethics committees and ethics experts often serve as a smokescreen for the biopolitical mechanisms and processes at work in many laboratories and companies. For CAE, ethics committees and programs too often remain complicit with the dominant logic of capitalism, a logic which silences other, less "productive" voices and interventions. The Federal Bureau of Investigation's initiation of the legal proceedings against Steve Kurtz, a CAE member, for alleged possession of biological materials with the intention of using these for the production of bioweapons in 2004,[28] serves as evidence that the aesthetics and ethics of revelation employed by contemporary bioartists have serious political effects, both within and without the art world.[29] While CAE can thus perhaps sometimes disappoint by rhetorically shaping their projects through philosophically weak, even conservative, notions such as "individual sovereignty" and "humanistic purposes," it could be argued that the ethical force of their work consists in creating the possibility of transforming the biopolitical power relations in the world, analyzing the temporary stabilizations of power, and making some of the agents occupying its nodal points accountable by raising the question of responsibility with and for them.

It needs to be acknowledged that artists are by no means uniquely predisposed to posing such a challenge and enacting such an alternative: similar tasks can be undertaken by philosophers, cultural and media theorists, media practitioners, or policy makers. And yet artists allow for these bioethical and biopolitical issues and concerns to circulate in a particular "poetic" language, whereby *poièsis* is understood, after Heidegger (1977), as a technological process of bringing forth, or creation. In other words, they are actively and purposefully engaged in the construction and circulation of alternative narratives about life (even if it cannot be guaranteed that these narratives will be received as interesting or aesthetically pleasing or that they will be noticed at all by the public). By narrative I mean here something stronger than just a "story about something," although it is that too.

But these narratives have a performative force, in the sense that they "actively shape what the technologies mean and what the scientific theories signify in cultural contexts,"[30] just as science fiction literature does in N. Katherine Hayles's account. For Hayles narrative is a more embodied form of discourse than a systemic abstraction aimed at eliminating the grain of the matter for the sake of a coherent theory. It can thus delve into specific cases—SymbioticA's *Pig Wings*, Stelarc's *Extra Ear*—through which what Stelarc calls "alternate body architectures," and what we may by extension call alternate living structures and environments, can be enacted in a multiplicity of singular ways. Even though bioart projects are often depicted through the rhetoric of panic and hysteria in mainstream media, I propose instead that experiments such as Kac's *GFP Bunny*, CAE's *Cult of the New Eve*, or Jeremijenko's *OneTree*[31] actually enact a more radical politics and ethics of life-as-we-perhaps-do-not-know-it-yet.

More specifically, I want to suggest that it is performativity at the ontological level of life (manifested in raising questions such as, What is life? What counts as human life? How are the boundaries of the human established and maintained? Is the distinction between humans, animals, and machines solid or arbitrary? Are differences between species a matter of grade or kind?) that bioart enacts most forcefully. Indeed, the performative aspect of the instability of the boundary between human and animal, as well as human and machine, which many bioart projects expose, provides a framework for an otherwise bioethical thinking. However, it is not the constant multiplication of differences and the infinite generation of alternatives that is most interesting for me here. It is only in the never receding obligation to address the question of the (other) human and nonhuman, and to come to terms with humans' "originary technicity," that these different projects will be truly ethical. As should be clear by now, bioethics, the way I am approaching it in this volume, does not stand for a mere enactment of a moral program which could then be securely applied to all cases of bioart. Rather than primarily deal with whether we *should* treat humans, animals, tissue, or life "like this"—that is, whether Kac "has the right" to make the bunny green, whether Stelarc is "responsible enough" in implanting an "unnecessary" ear on his arm (although these questions are important too)—bioethics as performed by bioartists interrogates its own relation to the discourses through which it is traditionally set up: the discourse of "informed consent," of the "net sum of happiness in the world," and of the "sanctity of life."[32] The question of the treatment of humans, animals, tissue, and "life" is certainly valid and significant, but in order to fully realize and responsibly respond to it, the very conditions and limitations of the possibility of its posing need to be investigated.

These issues are addressed most explicitly in the work of Adam Zaretsky—a self-declared "vivoartist" who is all for reinserting "fun for fun's sake into the social"[33] and for "things getting really weird."[34] He admits he would be keen to see some more "interesting mutations": "iridescent humans, spotted and striped, with multiple limbs."[35] Zaretsky here seems to take on a role of a Nietzschean philosopher, hitting at the established values around biotechnology and "life" with a double-edged hammer of serious play, in order to undertake their transvaluation.[36] His blasphemous-sounding proposal "for a new eugenic enhancement" needs to be read—ironically but also reflexively—within the context of the transformations that are already sanctioned in our culture and that are taking place on a day-to-day basis: cosmetic surgery, selective breeding of animals for food, the use of children's growth hormone. Zaretsky's work thus exposes and challenges the justification of the acceptable eugenic decisions we regularly make and the cultural and political values that underpin them. In a similar vein, Eduardo Kac situates his transgenic experiments with the GFP K-9, a genetically manipulated dog which served as a predecessor to the GFP Bunny, in the context of the long history of animal breeding and the earlier practices of genetic "engineering." Indeed, he suggests that the dog emerged only in its current "natural" form as a result of the early human-induced selective breeding of adult wolves that led to the retention of immature characteristics in the newly developed species.[37]

As we can see from the above, a number of bioartists are raising questions about the validity of the current moral panic regarding "eugenics," a panic which forecloses all too early the debate on life, its sociocultural valuations, and the historical practices involved in its legitimate and illegitimate manipulation. Although the effect of such questioning may seem shocking to some and irresponsible to others, artists such as Kac and Zaretsky skillfully highlight the constructedness of "nature" as well as the assumptions that accompany calls for its preservation. They also shift the debate from the generic concern about the alleged sanctity and untouchability of "life" in its transhistorical essence to the analysis of the power conditions involved in transforming this life in particular ways in different historical periods and geographical locations. Zaretsky's call to "critically embrace the processes of life's permanent and inheritable alteration" prepares the ground for such an analysis. However, even if mutation per se is something Zaretsky treats as both interesting and inevitable, he is worried about the commodification of life by the biotech industry as well as the narrow normative framework for what counts as "good life" developed in the business context.

His Workhorse Zoo installation, a collaboration with another artist, Julia Reodica,[38] shown at the Salina Art Center in Kansas in 2002, consisted of an eight-foot-square clean room filled with "typical scientific test subjects"—roundworms, mice, yeast, wheat plants, zebra fish, flies, frogs, and bacteria—as well as *Homo sapiens* (Zaretsky and Reodica themselves). The artists wore different costumes on each of the seven "theme days" during which the exhibit took place. All the species contained in the clean room were encouraged to interact with each other, an interaction that took the form of each fending for themselves (and thus inevitably eating each other) on the "Caveman Day." In an article about the installation Zaretsky positions Workhorse Zoo as carrying out "ethical edge" work—interrogating various "competing and ideologically inflexible concepts of animal cruelty and care" but also raising questions for those wanting to claim "a moral high ground" for their particular concepts and values.[39] Exposed to the eyes of the gallery and Internet audiences in a reality-TV-show manner, Zaretsky and Reodica's zoo was a knowing response to the public hunger for "real life stories." This time, however, the public were asked to reflect on and critically reevaluate the anthropocentric formulation of the idea of "life" that normally underpins TV addicts' hunger. Zaretsky himself had also prearranged a number of "bioethics quizzes," which he gave to visitors to the exhibition. The quizzes included the following questions:

What is your view on the Origin of these organisms, before domestication and now as mail-order commodities, particularly laboratory breeds [?] Where should they be if not where they are? Why is this your belief? . . .

What is your view on the Housing of these organisms, in particular the ethics of multispecies housing? Should multiple organisms be allowed to live together under the jurisdiction of human compatriots? Why is this your belief?

What is your view on the variety of settings collaged upon each other inside this education / entertainment / agitprop environment? How is this different than a nature-ish setting at a zoo or the minimum requirements for keeping laboratory animals? Is "The Wild" any better? . . . Why is this your belief? . . .

How well trained are you in judging artistic merit of independent, multispecies performance? Do you have any experience in art criticism or art history? Are you a bioethicist by trade? Not being a fan of expert knowledge, we ask, how do you decide what is real art, hollow art, farcical art or credible art and are those judgments mutually exclusive? How do you define what it means to be human, humane, good or just? Why is this Your belief? . . .

What is your opinion on the accidental witnessing of mouse cannibalism? It is not unusual for mice to eat each other but it was not planned for. In a lab situation it might be left underreported or filed away. . . . Do you blame the artists for this act or do you give the mice some agency for their own behaviors? How do you differentiate between human effect and animal instinct and/or animal consciousness? Are mice capable of being inhumane or inmousish? Is interspecies guilt a two way street? Why is this your belief?[40]

Zaretsky and Reodica's art project, and Zaretsky's own writings on biotechnology and art, can perhaps be described as "bioethics in action." Creating, with the help of the Workhorse Zoo, a thinking–living environment, the artists place the responsibility on the viewers not only to engage with an external art object but also to examine their own assumptions about, and their affective investments into, the ideas that underpin their reactions. By constantly reiterating "Why is this your belief?,"

Zaretsky can be said to attempt to denaturalize the "reaction" (be it disgust, abhorrence, passion, or indifference) to his unexpected zoological interactions and to visibly situate it in the conceptual framework that informs it. Interestingly, even though the artist seems to recognize the corporeal component of perception—the "gut feeling" of disgust, perhaps, that some viewers may experience when witnessing multispecies cannibalism in the Zoo—he insists on the need to connect this sense of being affected in a certain way with a theoretical and moral framework through which these affects can be described. By asking the viewers, "How well trained are you in judging artistic merit of independent, multispecies performance?," Zaretsky implies that a reaction is already a form of a decision, even if not always a conscious or well-thought-out one. Insisting on the need to combine the "gut reaction" with rational justification, Zaretsky also offers a way of out of the paralyzing moralism of conviction and a passage into a nonnormative ethics of responsibility.

It is precisely in troubling the distinction between the passivity of bodily affect and the activity of the decision that his bioethical ruminations become most interesting for me. We could perhaps go so far as to suggest that Zaretsky manages to weave together the Spinozian–Deleuzian understanding of ethics as "an *ethology* which, with regard to men and animals, in each case only considers their capacity for being affected"[41] with its Levinasian–Derridean conceptualization as openness to infinite alterity. The meeting point of these ethical positions takes place in the idea of—distributed but not suspended—agency, which here belongs to mice as much as it does to men. But, let me hasten to add, it does not belong to all species and life forms in exactly the same way. A good explication of the Deleuze-inflected notion of "distributive and composite agency" has been provided by Jane Bennett in her article, "The Agency of Assemblages and the North American Blackout." Bennett argues there for "the conceptual and empirical inadequacy of human-centered notions of agency" and proposes instead "a notion of agency that crosses the human–nonhuman divide."[42] She terms it "the agency of assemblages: the distinctive efficacy of a working whole made up, variously, of somatic, technological, cultural, and atmospheric elements."[43] One point that Bennett does not pursue too far in her analysis, though, is that agential differentiation within particular "assemblages"—complex events and phenomena such as the North American blackout that took place on August 14, 2003, or the Workhorse Zoo that was enacted in Kansas from January to March 2002—is important: different "actants" will play different roles within an assemblage at any particular time, and these roles will also have different effects.[44] Some of these will be negligible, while others will matter a lot. Also, not all parts of the assemblage will be affected in the same way at the same time. What is more, there also exists a very real possibility of the emergence of antagonism within an assemblage. This antagonism will be triggered by competing claims (or, to use less human-centric vocabulary, competing forces) between different actants: for example, mice, air, humans, bacteria.

In spite of agreeing with Bennett on the distributive character of any form of agency, including a moral one, I would like to suggest that partial assessment of the situation by the human, who is capable not only of recognizing in himself or herself this propensity for being affected but also of theorizing this propensity, constitutes a (necessarily shaky) foundation of what I have been referring to as "alternative" bioethics throughout this book. By saying this, I am not promoting here what Bennett terms "human exceptionalism," whereby the concept of agency entails a celebration of the distinctive power of human intentionality and a desire to elevate the human mode of being above all others,[45] even if I do support the idea of human (and nonhuman) singularity. Singularity stands for the recognition of the difference of various species, life forms, and inanimate objects that is never completely stabilized and that is only ever enacted *as difference* in a relation with others (other species, life forms, etc.) in a multiplicity of singular ways.

In the chapter "The Singularity of New Media" in his *Digitize This Book!*, Gary Hall draws on Derrida's work to explain the relationship between singularity, ethics, politics, and decision in the following terms:

> [T]he decision as to what politics is and what it means to act politically cannot be left completely open and incalculable. In fact, if we were to agree that politics is arbitrary, we would be returning to a conception of the sovereign who is always

instituted in an inherently unstable and irreducibly violent and arbitrary manner, but whose conditions of institution are usually forgotten or obscured. We would also be reinstating the autonomous and unified sovereign subject as the originator of any such decision and ethics. A decision that remained "purely and simply 'mine,'" that "would proceed only from me, by me, and would simply deploy the possibilities of a subjectivity that is mine" would not be a decision, according to Derrida. . . . It would be the "unfolding of an egological immanence, the autonomic and automatic deployment of predicates or possibilities proper to a subject, without the tearing rupture that should occur in every decision we call free."[46] . . .
So, we cannot devise a "coherent alternative" preconceived program or plan for founding a politics on ethics that is going to be unconditionally and universally applicable to every circumstance and situation. But, if we are not going to simply reinstate the autonomous and self-contained subject, or be determined purely by the pragmatic demands of the particular context, neither can the decision regarding politics be entirely arbitrary. We have to base it on *universal* values of infinite justice and responsibility; there has to be *some* link between ethics and politics. Once again, we find that making a just and responsible decision for Derrida requires respect for *both* poles; we just have to go through the trial of taking a decision in such an undecidable terrain. And it is here, too, "that responsibilities are to be re-evaluated at each moment, according to concrete situations" . . . ,[47] because each time a decision is taken is different, each situation and context in which a decision is taken is different. So on each occasion we take a decision we have to invent a new rule, norm or convention for taking it that exists in a relation of singularity to both the infinite demand for *universal* values of justice and responsibility that is placed on us by the other (by the other in us), and each particular, finite, "concrete" conjunction of the "here" and "now" in which this demand occurs.[48]

We can see from the above that both Derrida and Hall remain far from celebrating the superiority of the human; they are also clearly suspicious about the stability and cohesion of "human intentionality" which could provide a foundation for ethics. Indeed, it is in the tension between the lack of such stability and certainty on the one hand, and the need to act in this unstable territory on the other, that ethics and politics take place—and they take place always differently, always in a singular way.

Significantly, Bennett also explains that by focusing on what she terms "the cascade of becomings," she is not denying intentionality and its force altogether but rather seeing intentionality as less definitive of outcomes.[49] Not only is a decision, it seems, "made in an undecidable terrain":[50] its outcomes are also undecided. These outcomes are never solely a consequence of the human act of decision making, even if they have been triggered by it. The difference between the Derrida–Hall position and the Deleuze–Bennett one is that the former focuses on the (necessarily unstable, compromised, and perhaps even rare) instances of human decision amidst all the different material events, connections, and occurrences which are constantly taking place in the world, while for the latter these instances are not so relevant or interesting amidst the whole "cascade of becomings." Even though Bennett herself reneges on the notion of "full responsibility" that can allegedly be borne by humans for material effects in the world, she makes a very Derridean point when she suggests that "Perhaps the responsibility of individual humans may reside most significantly in one's response to the assemblages in which one finds oneself participating—do I attempt to extricate myself from assemblages whose trajectory is likely to do harm? Do I enter into the proximity of assemblages whose conglomerate effectivity tends towards the enactment of nobler ends?"[51] Intriguingly, while the concept of "harm" can be easily explained through the Spinozian–Deleuzian notion of "goodness," whereby "The good is when a body directly compounds its relation with ours, and, with all or part of its power, increases ours,"[52] the idea of "noble ends" clearly brings back theorization, rationality, and reflexivity, limited and partial as they are.

I am therefore not advocating a return to an unreconstructed humanism with the proposition that partial assessment of the situation by the human, who is capable not only of recognizing in himself or herself this propensity for being affected but also of theorizing this propensity, should constitute a foundation of any "alternative" bioethics. I am only suggesting that a reflexive moment in which an evaluation is to be made about the powerfulness of an encounter between particular elements–bodies, and about the supposed "goodness" of this encounter, is necessary for ethics to take place. This proposition constitutes a supplement to the Spinozian concept of "goodness." To judge this increase in our power or even to actually understand who is the "we" in this relation, a moment of at least temporary differentiation between different parts of the assemblage has to occur. If ethics is not to turn merely into a formal exercise in the study of life flow's increase and decrease, one needs to be able to determine why some bodily encounters *matter* more than others, and who they matter to at any particular time. One must also be able to answer Zaretsky's question "Why is this your belief?" even if we are to concede that this answer will only ever be provisional and partial.

The question "Why is this your belief?," which punctuates Zaretsky's bioethics quizzes, therefore entails a call to "the human" (the art gallery visitor, the meat eater, the lab scientist) to take responsibility for life and to make decisions—informed, reflexive, but also often inevitably violent—about its different processes. It is precisely the recognition of the inevitability of violence that makes Zaretsky's open-ended bioethical project so different from the humanist bioethics of "informed consent." Indeed, his question also inheres a suspicion toward an illusion about the human as a self-contained moral agent, fully accountable for his moral actions and political agency. The Zoo Workhorse shows that a straightforward normative valuation becomes problematic in a network from which the human/the artist does not disappear as an agent altogether but in which agency becomes more distributed, composite, and networked.[53] This is, as Bennett puts it, an "agency that includes nonhumans with which we join forces or vie for control."[54] This model challenges the more traditional social sciences conviction that "agency attaches exclusively to persons and [that] social structures 'act' only as they *thwart* human agency."[55]

The Extra Ear of the Other: On Being-in-Difference

The issue of distributed, human–nonhuman agency is addressed most interestingly in the work of the Australian performance artist Stelarc. I have written elsewhere about the ethical aspects of Stelarc's earlier robotic and prosthetic projects,[56] but in the final part of this chapter I want to focus on his more recent experiments in the area of bioart, which involve engagement with soft tissue and plastic surgery techniques. Stelarc's *Extra Ear: Ear on Arm* project (2006–2007) provides a focal point for my closing remarks on ethics, responsibility, agency, and decision in bioart.

The ear, as Jacques Derrida poignantly observes, is uncanny: it is double in more than one sense. First of all, it is always that of the other (ear). The ear can also be open and closed at the same time. The only organ that cannot voluntarily shut itself, it always remains ready to hear, to receive, even if its "owner" is not actively listening. The very possibility of speaking and writing and, thus, of communicating, exchanging, and, more generally, of being with others comes from the ear.[57] But what happens when the ear wanders down from the side of one's head to one's arm—as it does in Stelarc's *Extra Ear* project—and when it mutates from a receiving to a transmitting organ? Stelarc's *Extra Ear* mimics the actual ear in shape and external structure, but rather than merely hearing, it will wirelessly transmit sounds to the Internet, thus becoming a remote listening device.[58] The *Extra Ear* inevitably draws in the observer's eye; it attracts a curious gaze while also encouraging a cross-sensual exchange between bodies and organs which goes beyond the functionalism of information exchange. But what is Stelarc trying to *say* with this *Extra Ear*? And what would it mean to really h-*ear* him, and respond to him?

Stelarc's recent visual and aural performances shift bodily architecture into the hybrid terrain of walking heads, hearing arms, and fluid flesh (*Walking Head Robot*, 2006; *Blender*, 2005—with Nina Sellars). The invitation extended to his audience to open up to his reconstructive and often invasive bodily projects at times seems to fall on deaf ears, which is why some responses to his work end up in a solipsistic, and often moralistic, position of *not hearing him at all*, and of deciding in advance what he is trying to say and why it is wrong. The artist's provocative statements that "the body is obsolete" and that we need to devise alternate anatomical architectures which are not emerging in evolution but which become possible through engineering[59] are often interpreted as simply meaning that the body is both inadequate and unnecessary, that it is only an obstacle in our "technological age." The French philosopher Paul Virilio even went so far as to accuse Stelarc of enacting a new version of "clinical voyeurism": a dangerous strategy aimed at improving the human but ultimately signaling, for Virilio, both the end of art and the end of humanity.[60] So what does it mean to really *hear* Stelarc and *respond* to him? How can, or should, we engage with his *Stomach Sculpture* (1993), *Partial Head* (2006), and *Extra Ear*? On what grounds can Stelarc's projects themselves be described as *respons*-ible (or not)? What values does his reworking of the architecture of the body challenge, and does this challenge imply any broader epistemic transformation?

Let me put forward two ontological propositions which have already made a number of appearances in this book and which will provide a context for my reading of Stelarc's artwork—from his older experiments with suspensions and prosthetics to his more recent engagements with artificial intelligence, biomaterial, and plastic surgery:

1. The body is not "inadequate" in any prescriptive sense, nor has it suddenly become obsolete in what is called "the technological" or "the new media age"—even if we consider the body as contingent and epiphenomenal, rather than as an end product of intelligent design or divine creator.
2. The human has always been technological; or, to put it differently, technology is what makes us human.

To consider the body an "inadequate" structure in a world of information flows, machinic wars, and nanotechnology does not therefore amount to a melancholic renunciation of the fantasy of human potency. We can better understand this statement as a pragmatic recognition of humans' dependency on other life forms as well as inanimate objects. Rather than reduce them to a naive prophecy of a postflesh world in which man will eventually overcome his technological limitations, Stelarc's performances should rather be seen and encountered as taking on, and making both visible and audible, the differential relationship the human has always had with technology. In other words, they show technology as constitutive of the idea and materiality of both "the human" and "the body."

We could thus perhaps suggest that Stelarc's work enacts what Bernard Stiegler has named "originary technicity," which I have referred to earlier on in the book. In *Technics and Time*, Stiegler denaturalizes the position of "nature" as primary and of man as originally both savage and existing in unity with nature. "Pure nature" is shown to be a logical impossibility, as for man to exist, he needs to differentiate himself—through what Stiegler calls a technical accident—from the world, from "nature," from what is nonhuman. I would argue that Stelarc helps us grasp this new understanding of the relation between human and technics, in which a technical "object" (say, a computer or a cable) cannot be just a utensil. He shows us that there is no simple end–means relation between "man" and "his" technologies.[61] Instead, the "nature" of the human is produced only *as* and *through* technicity. Hominization, as Stiegler explains, that is the emergence or "production" of the human, is a moment of rupture, a process of exteriorization in which the skeletal being rises on two feet and reaches for what is outside him (tools, language).[62] This process of exteriorization marks the appearance of the technical. "Or again: the human invents himself in the technical by inventing the tool. . . ."[63] The moment of technicity, a "putting-out-of-range-of-oneself," is a moment in which the (not yet formed) self reaches for something: an alterity it does not know.[64] In Stelarc's artistic performances—from the earlier suspension events where technology came in the form of hooks and ropes, to the more recent robotic and artificial intelligence experiments such as *The Walking Head Robot* and *The Prosthetic Head* (2003)—the human is positioned as being born out of this relationship with *tekhnē*. This "technical condition" has always organized our being in the world, but it can be argued that the age of new technologies and new media has led to an acceleration and intensification of this condition. It has also prompted a rethinking of the more traditional subject–object based models of the relationship between "humans" and "technology," "us" and "the world."

Stelarc thus seems to provide us, in his own way, with a radically different perspective on technology. He departs from the dominant Aristotelian model of *tekhne¯* in Western societies, where technology is principally seen as a tool for the human, and outlines instead a more systemic and networked model of human–nonhuman relationality, in which prostheses function as structuring devices rather than unwelcome additions to the already complete molecular entities. For Stelarc technology is first of all an environment, a network of forces and relations, rather than merely an object or simply a "human other." Brian Massumi makes a similar observation in relation to Stelarc's earlier works, including *Ping Body*, *Fractal Flesh*, and *The Third Hand*, when he suggests that the body and its objects, ropes, goggles, computers are prostheses *of each other* and that matter itself is prosthetic.[65] Massumi goes on to maintain that "the body is always already obsolete, has been obsolete an infinity of times, and will be obsolete countless more—as many times as there are adaptations and inventions. The body's obsolescence is the condition of change."[66] Obsolescence is thus here a positive driving force, a promise of transformation. Massumi proposes we approach it as an opening rather than a source of anxiety: first, because all these adjustments of the functioning of the human body and its combinatoric possibilities are inevitable, and, second, because every adjustment implies on some level "an interruption of the old functioning to make an opening for the new." Stelarc's aforementioned pronouncements on the obsolescence of the body and the need for its augmentation, with all the connotations of techno-economic progress and social evolution they entail, can thus best be heard and interpreted, I want to suggest, as provocations that are testing the limits of the technodeterministic discourse, a discourse in which technology is positioned as determining the way the world develops. They reveal the possibility, or even inevitability, of the unpredictable, the accidental, and the unknown in any developmental trajectory.

Similarly, for Jane Goodall, there is no dramatic confrontation between the body and the machinic other in Stelarc's projects: she sees them instead as attempts to harmonize organic and technological components in diverse and nuanced ways.[67] Since in Stelarc's universe of multilateral and synthetic fusion "everything is interwoven with everything else,"[68] the claim to human superiority is undermined on the material level by the multiplicity of crisscrossing connections within the organic–mechanic system, in which "human" is only one possible, and unstable, nodal point. However, Stelarc's artistic experiments do more than just explore the complex connectivity of the system, I would argue. They also foreground, in a playful but also cautionary way, the possibility, or even the inevitability, of an "accident" within it. It is precisely through art as accident—broadly devised by the artist together with teams of engineers, computer programmers, and surgeons but ultimately left "indecisive," allowed to take on "a life of their own"—that we enter in Stelarc's work the realm of what could be described as creative evolution. Drawing on the work of Henri Bergson, Sarah Kember explains that the concept of creative evolution lacks the finality and teleology associated with the

Darwinian idea of evolution, in which nature is seen as a mathematical apparatus ruled by predictable laws. She sees the overall adoption of this concept in both biological and social sciences as politically conservative, limiting change in the political realm to the allegedly fixed algorithms to be detected in nature, and envisaging social change as always already foreseeable. Kember instead foregrounds the "principle of contingency," which she sees as fundamental to Bergson's concept of evolution and which she also comprehends as more enabling politically. She argues that "the value of contingency, of not knowing what life can do, lies initially in its exposure of the illusion of determinacy. . . . [I]t indicates a change in ways of being, knowing (including scientific knowing), and doing, which become based less on the mastery by certain subjects of their chosen objects and more on inter- or rather 'intra'-subjective relationalities."[69]

Human agency does not disappear altogether from this zone of creative and contingent evolution, but it is distributed throughout a system of forces, institutions, bodies, and nodal points. This acknowledgement of agential distribution—a paradox that requires a temporarily rational and self-present self which is to undertake this realization—allows for an enactment of a more hospitable relationship to technology than the paranoid fear of the alien other, manifested in the current moral hysteria about human cloning, body modification, and biosurveillance. In its embracing of the delay, the failure, and the accident, this hospitality—an ethical disposition in which the self recognizes itself as always already "invaded"—also avoids the uncritical fascination with "being wired," networked, cyberspaced. Indeed, Stelarc talks about adopting "the posture of indifference" in relation to his work; about not attempting to entirely control the event, about letting it happen after starting it off. This pronouncement reveals a tension between the modernist notion of artist as lone creator and instigator of ideas, on the one hand, and the cybernetics-informed understanding of artist as a node in the network of exchange on the other. However, Stelarc's "posture of indifference" can also be read as being "in-difference," as what Derrida would call, after Emmanuel Levinas, hospitality toward infinite alterity, an opening of oneself to what is not in one. Naturally, the decision about adopting this posture of indifference, about not having any expectations, is made from a temporarily stabilized point of human agency. Still, we should perhaps read it as not *just* a rational decision, but also as bodily passivity, as letting oneself be-together-with-difference, with-technology. Again, we are not talking here about the pairing of "human" and "technology" seen as separate entities but rather about human agency and corporeality as being always already reliant on, connected to, and becoming with, *tekhnē*.

Stelarc repeatedly tells us that he has no ambitions to be a philosopher or a political theorist. He refuses to be prescriptive in his work and so will not instruct us as to how we should treat our bodies or how we should coexist with technology. However, listening to Stelarc will allow us to envisage a more effective politics and ethics. This will be a technopolitics of distributed agency and suspended command, informed by an ethics of infinite—and at times crazy, shocking, and excessive—hospitality toward the alterity of technology (that is always already part of us). This is not to say that "we are machines" or that "we are all Stelarcs now," as the Krokers put it.[70] If humans and machines are collapsed—as they tend to be in some current accounts of "the network society"—into a fluid epistemology in which difference is overcome for the sake of horizontal affective politics, it is not clear any more who encounters, responds to, and is responsible for whom in an ethical encounter and, what is more important, why it should matter at all. We therefore need to retain, as I argued earlier on in this chapter, this idea of the human (and that of the body), even if we must place it in suspension. The narrative of seamless coevolution between different biological and cultural entities that depicts the whole world as "connected," it seems to me, threatens to overlook too many points of temporary stabilization that have a strategic political significance. It is via these "points of temporary stabilization" that partial decisions are being made, connections between bodies are being established, aesthetic and political transformation is being achieved, and power is taking effect over different parts of "the network" in a differential manner. At the same time, "the human" should not be seen as an essential value or a fixed identity but rather as a strategic quasi-transcendental point of entry into debates on agency, human–technology environment, politics, and ethics. Ethical thinking in terms of originary technicity that brings forth the very concept of the human—as an ethical response to alterity—allows us to affirm the significance of the question of responsibility without needing to rely on the preestablished value system that legislates it or resort to an unreconstructed a priori moralism. This is one nondidactic lesson we can learn from Stelarc's posture of in-difference. And this Stelarc-inspired "being-in-difference" can perhaps serve as a pointer for outlining a bioethics for humans, animals, and machines in the age of new media.

Conclusion

Nonclinical Hospitality and "The Human Becoming"

A study of the academic and professional discourse known as "bioethics" that has framed and legislated the debates on life and its technological mediations, this book is also an attempt to develop a different framework for thinking about ethics and life in the age of new technologies and new media. The alternative bioethics proposed here goes beyond the institutionalized paradigm of the doctor–patient relationship. Under the current biopolitical regime, we are all positioned as patients, with our lives, both in their biological and social aspects, becoming an object of explicit political and commercial interest. My principal concern in the book has therefore been to take the debate on bioethics beyond its traditional homes of analytical philosophy and disciplines related to medicine and into those fields where questions of the human and human life have been addressed from a different angle, allowing for a more explicit consideration of the issues of meaning, mediation, and politicization: media and cultural studies, cultural theory, and the arts. In doing so, I have looked at a number of conceptual and material transformations that the key terms of traditional bioethics such as the human, the animal, and "life itself" have undergone over the recent decades.

The "new" bioethics I have attempted to sketch throughout the course of this volume, with the help of selected philosophers of life and cultural theorists, takes the form of a conceptual framework or a set of nonnormative ideas which can only be considered and enacted in specific instances. Its driving force comes from a content-free obligation toward other beings and forms of life, some of which we do not even notice, do not comprehend, or are unable to name. Bioethics as an ethics of life understood in this way is nevertheless different from an ecological ethics that cares about the whole universe: it is premised instead on a recognition that singular, and often violent, decisions will need to be taken as to how to relate to this multiple "difference of life." This form of bioethics will therefore not speak *against abortion* or *for vegetarianism* in advance, on the basis of a general and prior "respect for life," because no such blanket moral injunction can be deduced from this notion of obligation. This infinite obligation to respond to life does not mean that one has to respect carrots, fetuses, or cyborgs upfront but rather that we need to consider their shared genealogy and genetic heritage, without the immediate application of the anthropocentric principle that allows us to see humans as not just superior to other beings but also as fully knowable and uniquely cognizant.

The book does not by any means advocate a total rejection of the existent bioethical tradition, developed by moral philosophers and clinicians and applied in hospitals and medical research institutes. However, the birthplace of the bioethics proposed here is at the very margins of the dominant system of *hospital*ity, thus positioning it as this system's not always acknowledged exception. Open, in principle, "*to who or what turns* up, before any determination, before any anticipation, before any identification, . . . whether or not the new arrival is the citizen of another country, a human, animal or divine creature, a living or dead thing, male or female,"[1] this alternative bioethics also takes as its task an examination of the historical formation and ideological structuration of "the human," and of many of the concepts positioned as the human's "other," such as the animal and the machine. The human does not disappear from the kind of nonhumanist bioethics envisaged here: in fact, it functions as its strategic point of entry. What we are dealing with, however, is not so much a "human being" understood as a discrete and disembodied moral unity but rather a "human becoming": relational, co-emerging with technology, materially implicated in sociocultural networks, and kin to other life forms.

The philosopher Bernard Stiegler, whose arguments resonate strongly in this volume, revisits early paleontological narratives with a view to presenting a different history of the hominoid as always already technological, that is, as emerging in-relation-to, and in-difference-with, his tools. However, a decision as to the status of the human and his or her kinship with other moral subjects, as well as objects, still has to be constantly taken and retaken—and even more so in the present age of new media developments and biotechnological experiments. This does not mean that there was a point in time when such a decision could be seen as unnecessary or self-evident but rather that the current speed and intensity of technological mediations requires a new modulation of this process of decision making. In other words, even if this book is based on the premise that the human is always already other—a premise supported both by Emmanuel Levinas's pragmatic transcendentalism, which seeks the origin of human subjectivity in the infinity of the other, and by Stiegler's technological materialism, which perceives the human as emerging via his tools—it nevertheless asserts the need to focus on the particular events associated with our current "technoculture" and the way new technologies and new media affect and alter the formation of the human.

By pointing, with Levinas and Derrida, to a place of foundational difference as the origin of a bioethical injunction, the nonhumanist bioethics outlined here challenges the hierarchical system of descent through which relations between species and life forms have traditionally been thought. Informed by the need for a decision, always to be made anew, about what to do, this bioethics of alternative hospitality assumes, or rather accepts, responsibility for the lives and deaths of multiple, human and nonhuman, others. The concept of responsibility—in the sense of an *exposition* and inevitable *response* to the alterity of other becomings and life forms—is fundamental to this particular ethical framework. However, such responsibility is much less frequent and much more difficult to enact with any certitude or moral uprightness. Indeed, as Francisco Varela acknowledges in *Ethical Know-How*, many decisions which we consider ethical are in fact only spontaneous reactions to the goals that present themselves before us. But while cyberneticists and cognitive psychologists interpret such reactions as ethical events, while also positing the smooth functioning of a semantically neutral system as an ethical value, for me ethics occurs when the very processes of reaction and feedback come to a temporary halt, when the system unworks itself, when a ghost enters the machine to disrupt its calculation.

Ethical responsibility thus needs to be distinguished from instant pragmatism, a quick semi-intuitive reaction in a given set of circumstances. Nor should it be reduced to mere relativism, a contextual decision undertaken by an individual according to his or her own moral criteria (or mere whim). Even though singular contexts and instances are important when it comes to ethical decisions, we must not forget about the broader horizon of responsibility which is also that of an ethical demand: one that reminds us that we are already indebted to the other—which is perhaps another way of saying that we are connected, relational, and hence dependent on what is before us, in both a temporal and ethical sense.

This hospitable but also inevitably violent bioethics of responsibility for "life itself" inheres a double injunction: to stabilize and to transcendentalize. Both actions will be temporary and strategic. We need to stabilize in order to manage the "flow of life" and conceptually carve out entities, such as "the human," "the other," or "the virus," if we want to consider how the competing claims for responsibility can be resolved in particular circumstances and whether all forms of relationality matter in exactly the same way at a given moment. We also need to transcendentalize, in the sense that particular bioethical claims, problems, and issues (abortion, cloning, the production and distribution of GM foods or of formula milk) need to be temporarily isolated from the whole complex network of sociopolitical circumstances and elevated over other problems in order to be considered at all. However, it is the wider commitment to responsibility—something that for me amounts to the acknowledgement of the lived materiality of other "becomings" more than to a metaphysical injunction—that differentiate this ethics from mere individualistic decisionism.

Since most moral decisions are just reactions, spontaneous readjustments within the dynamic system of forces, "true" bioethical events are therefore rather rare. This does not make the need for a bioethics in the age of new media any less significant. Bioethics, as I have envisaged it in this volume, is premised on a long-term nonnormative commitment to understanding the production and emergence of what we call "life" in its multiple materializations, mediations, and symbolizations. It is also a way of cutting through the "flow of life" with a double-edged sword of productive power and infinite responsibility.

Notes

Preface

1. Andreas Broeckmann, "Introduction," Catalogue for *BASICS: Transmediale.05: International Media Art Festival Berlin* (2005), 1.

2. Apart from the work on biotechnology and ethics by a number of Deleuze-inspired writers (some of whom are listed in note 6 below), we should also recognize that attempts to raise bioethical questions have been forthcoming from other theoretical perspectives, situated outside bioethics' more conventional homes of analytical philosophy and the health care professions. We can mention here the following titles: Rosalyn Diprose, "A 'Genethics' That Makes Sense," in *Biopolitics: A Feminist and Ecological Reader on Biotechnology*, eds. Vandana Shiva and Ingunn Moser (London and Atlantic Highlands, NJ: Third World Network and Zed Books, 1995); Carl Elliott, *A Philosophical Disease: Bioethics, Culture, and Identity* (New York and London: Routledge, 1999); Sarah Franklin, Celia Lury, Jackie Stacey, *Global Nature, Global Culture* (London, Thousand Oaks, New Delhi: Sage, 2000); Donna Haraway, *The Companion Species Manifesto: Dogs, People, and Significant Otherness* (Chicago: Prickly Paradigm Press, 2003); or Andy Miah, *Genetically Modified Athletes: Biomedical Ethics, Gene Doping and Sport* (London and New York: Routledge, 2004).

3. Martin Lister et al., *New Media: A Critical Introduction* (London and New York: Routledge, 2003), 10; emphasis added. Lister et al. prefer the term "new media" to "digital media" or "electronic media" because of the purely technical and formal emphasis in the latter monikers. They also criticize the label "interactive media" for focusing exclusively on only one randomly chosen characteristic.

4. Wendy Hui Kyong Chun, one of the editors of the anthology *New Media, Old Media* (New York and London: Routledge, 2006), is highly critical of certain overoptimistic and one-sided adoptions of the "new media" term. She writes in the Introduction to the volume, "The moment one accepts new media, one is firmly located within a technological progressivism that thrives on obsolescence and that prevents active thinking about technology–knowledge–power" (9). And

yet she does not recommend abandoning the term altogether. Instead, she recognizes that "new media" has already been consolidated into a field with its own emerging canon and institutional space. However, Chun argues strongly against perpetuating the myth of the singular uniqueness of new media, insisting instead that the new "contains within itself repetition" (3). It is very much with similar concerns at heart that I am adopting this term in my book.

5. See Mark Hansen, *New Philosophy for New Media* (Cambridge and London: MIT Press, 2004) and Gary Hall, *Digitize This Book! The Politics of New Media, or Why We Need Open Access Now* (Minneapolis: University of Minnesota Press, 2008).

6. See Rosi Braidotti, *Transpositions: On Nomadic Ethics* (Cambridge: Polity Press, 2006); Adrian Mackenzie, "'God Has No Allergies': Immanent Ethics and the Simulacra of the Immune System," *Postmodern Culture*, vol. 6, no. 2 (January, 1996); Eugene Thacker, *Biomedia* (Minneapolis and London: University of Minnesota Press, 2004). Even though Mackenzie's essay opens with an address to Levinas, it moves on to embrace Spinoza- and Deleuze-inspired understanding of ethics as "a typology of immanent modes of existence."

7. In order to emphasize this point, the final chapter of the book includes a direct comparison between these two different traditions—and an explanation as to why I have chosen to work with the Levinas–Derrida–Stiegler triad.

8. In his reading of Martin Heidegger's "The Question Concerning Technology," Samuel Weber suggests that technology and nature are both processes of *poièsis* (creation). Weber here moves beyond the commonplace understanding of technology as relating to contemporary advances in media, communications, and medicine to embrace its etymological meaning of "technique, craft, or skill." If "the innermost principle of 'nature' is its impulse to open itself to the exterior, to alterity," technology-as-a-technique has to be seen as one form of nature, or even described as "more natural than nature itself," he argues. In Samuel Weber, *Mass Mediauras: Form, Technics, Media* (Stanford: Stanford University Press, 1996), 67.

9. Bernadette Wegenstein, *Getting under the Skin: The Body and Media Theory* (Cambridge and London: MIT Press, 2006), 80.

10. Wegenstein, *Getting under the Skin*, 29.

11. Helga Kuhse and Peter Singer, "Introduction," in *Bioethics: An Anthology*, eds. Helga Kuhse and Peter Singer (Oxford: Blackwell, 1999), 1.

Chapter 1

1. All these examples have featured as news items in different U.K. newspapers: "Immigrants Must Be Tested to Stop AIDS," *The Sun*, November 28, 2003; "Dr Dolly to Clone Human," *The Sun*, February 9, 2005; "Mental Health Campaign: Susan Begged for Help, Instead She Was sent to Prison," *The Independent*, April 29, 2007.

2. For more on this issue, see Carl Elliott, "Not-So-Public Relations: How the Drug Industry Is Branding Bioethics," *Slate*, December 15, 2003.

3. Fernando Cascais argues that the North American tradition of bioethics has been dominated by an adherence to moral principles rooted in utilitarianism and individualism, while the European tradition has been more influenced by virtue ethics and the idea of the common good. He states: "Ultimately, the leading determinant of the differences between North American and European bioethics lies in the fact that the thought that shapes the former is akin to the long-lasting tradition of economic liberalism, philosophical empiricism and pragmatism, sociopolitical individualism and ethic utilitarianism in the United States, meanwhile the latter is strongly conditioned by the omnipresence of the Welfare State, granted that beneficence and justice fulfill in Europe the theoretical and practical range that in the United States is attributed to autonomy. Besides, North American and North European bioethics are strongly tied to the tradition of the Protestant *Berufsethik*, described by Max Weber, while Southern Europe, Catholic and Orthodox, embody a long tradition of charitable health care that greatly restrains the boundary of a laic bioethics. Whereas in the United States the distinction between 'bioethics' in general and 'religious bioethics' ('Christian bioethics,' 'Jewish bioethics,' etc.) is clear, the latter expressing the distinct positions of various confessional morals, in Europe, especially in the South, the straight and plain impoundment of bioethics by religious morals is notorious. . . ." (29)

While this provides an interesting overview of the historical developments of bioethics, I suggest that the three different traditions—utilitarianism, deontological ethics, and virtue ethics—are at work, and often at loggerheads, in both North America and Europe. Fernando Cascais, "Bioethics: A Tentative Balance," *Studia Bioethica*, vol. 1, no. 1 (December, 2003): 25–34.

4. Helga Kuhse and Peter Singer, "Introduction," in *Bioethics: An Anthology*, eds. Helga Kuhse and Peter Singer (Oxford: Blackwell, 1999).

5. See Richard Kearney and Mara Rainwater (eds.), *The Continental Philosophy Reader* (New York and London: Routledge, 2005).

6. As examples of the recent "ethical turn" in the humanities and social sciences, one can cite the following volumes: David Campbell and Michael J. Shapiro (eds.) *Moral Spaces: Rethinking Ethics and World Politics* (Minneapolis: University of Minnesota Press, 1999); Marjorie Garber et al. (eds.) *The Turn to Ethics* (New York and London: Routledge, 2000), Howard Marchitello (ed.) *What Happens to History: The Renewal of Ethics in Contemporary Thought* (New York and London: Routledge, 2001), or Todd F. Davis and Kenneth Womack (eds.) *Mapping the Ethical Turn: A Reader in Ethics, Culture, and Literary Theory* (Charlotesville: University of Virginia Press, 2001), as well as my own, *The Ethics of Cultural Studies* (London and New York: Continuum, 2005).

7. However, it should become clear in time that a rigid opposition between these two different models—deontological and nonsystemic—is not entirely

sustainable. As a quick example, we could look at Kantian ethics and the way it is used by both "camps" as actually establishing a bridge between them.

8. Kuhse and Singer, "Introduction," 1. Chapter 2 engages with Van Potter's conceptualization of ethics in more detail. It investigates what bioethics would look like today if it had followed Potter's legacy.

9. Kuhse and Singer, "Introduction," 1.

10. Kuhse and Singer, "Introduction," 3.

11. Quoted in Onora O'Neill, "Kantian Ethics," in *A Companion to Ethics*, ed. Peter Singer (Oxford: Blackwell, 1993), 178.

12. Kuhse and Singer, "Introduction," 4.

13. Kuhse and Singer, "Introduction," 3. The introductory material on bioethics contained in the first part of the section " 'Traditional' Bioethics and Its Discontents" has been borrowed from my earlier book, *The Ethics of Cultural Studies*, in which I addressed the problem of bioethics for the first time.

14. Stephen Holland, *Bioethics: A Philosophical Introduction* (Cambridge: Polity, 2003), 4.

15. Catherine Waldby, *The Visible Human Project: Informatic Bodies and Posthuman Medicine* (London and New York: Routledge, 2000), 37.

16. See Judith Jarvis Thomson, "A Defense of Abortion," in *Bioethics: An Anthology*.

17. See Don Marquis, "Why Abortion Is Immoral," in *Bioethics: An Anthology*.

18. John Finnis, "Abortion and Health Care Ethics," in *Bioethics: An Anthology*, 13.

19. Finnis, "Abortion and Health Care Ethics," 14.

20. Finnis, "Abortion and Health Care Ethics," 19.

21. Michael Tooley, "Abortion and Infanticide," in *Bioethics: An Anthology*, 21.

22. Tooley, "Abortion and Infanticide," 23.

23. For a critique of the humanist assumptions behind the alife discourse and project, see Sarah Kember, *Cyberfeminism and Artificial Life* (London and New York: Routledge, 2003).

24. Albert R. Jonsen, *The Birth of Bioethics* (New York and Oxford: Oxford University Press), 241.

25. Tooley, "Abortion and Infanticide," 24.

26. Peter Singer, *Rethinking Life and Death: The Collapse of Our Traditional Ethics* (Oxford: Oxford University Press, 1994), 182.

27. For a critique of bioethics from a political perspective, see Donna J. Haraway, *Simians, Cyborgs and Women: the Reinvention of Nature* (London: Free Association Books, 1991), 44.

28. This is not to say that Singer is unaware of, or uninterested in, wider political issues: his numerous books, including *The President of Good and Evil: The Ethics of George W. Bush* (London: Granta, 2004) and *Eating: What We Eat and Why It Matters* (London: Arrow Books, 2006, coauthored with Jim Manson), testify to the contrary. However, the utilitarian model of moral philosophy he bases his ethical propositions on takes *the individual* as its subject and object, with singular ethical dilemmas often positioned as if they somehow functioned outside the broader political context.

29. John Harris, *Enhancing Evolution: The Ethical Case for Making Better People* (Princeton and Oxford: Princeton University Press, 2007), 2, 5, 8.

30. For more on the idea of "originary prosthesis" see my chapter, " 'The Future . . . Is Monstrous': Prosthetics as Ethics," in *The Cyborg Experiments: The Extensions of the Body in the Media Age*, ed. Joanna Zylinska (London and New York: Continuum, 2002).

31. Laura M. Purdy, "Are Pregnant Women Fetal Containers?," in *Bioethics: An Anthology*, 71.

32. See the FAB Web site: http://www.fabnet.org/

33. Jonsen, *The Birth of Bioethics*, vii.

34. Cited in Jonsen, *The Birth of Bioethics*, 27.

35. Two other documents that are considered crucial in the formalization of bioethics internationally are the United Nations Universal Declaration of Human Rights, proposed in 1948, and the 1964 Declaration of Helsinki (updated every four years).

36. Nikolas Rose, "The Politics of Life Itself," *Theory, Culture and Society*, vol. 18, no. 6 (2001): 1–30, 2.

37. http://www.pbs.org/wgbh/nova/doctors/oath.html

This Web site contains the text of both the ancient and the modern version of the Hippocratic oath.

38. The U.S.-based President's Council on Bioethics (originally called the National Commission for the Protection of Human Subjects of Biomedical and Behavioral Research) is one example of such institutions. See Jonsen, *The Birth of Bioethics*.

39. Cascais, "Bioethics: A Tentative Balance," 25–34.

40. Jonsen, *The Birth of Bioethics*, 333.

41. Warren Thomas Reich, "The Word 'Bioethics': The Struggle over Its Earliest Meanings," *Kennedy Institute of Ethics Journal*, vol. 5, no. 1 (1995): 19–34, 22.

42. Jonsen, *The Birth of Bioethics*, 372.

43. Andy Miah suggests: "If we are to give the notion of 'expert' any value at all—and there do seem good reasons to value experts, providing the nature of the dialogue with which expertise is communicated is not authoritarian—then there is a need for expert commentary on morality and communication to clarify

what is at stake for the public when new science stories break. Yet, when considering the possibilities of an ethical debate on science, asking who should be the expert on the morality of science further complicates the relationship between scientists and non-scientists," 411. Miah goes on to argue that we need to distinguish between scientific and ethics experts, and warns against expecting the former to do the job of the latter. Andy Miah, "Genetics, Cyberspace and Bioethics: Why Not a Public Engagement with Ethics?," *Public Understanding of Science*, vol. 14, no. 4 (2005): 409–21.

44. Miah, "Genetics, Cyberspace and Bioethics," 410.

45. Carl Elliott, *A Philosophical Disease: Bioethics, Culture and Identity* (New York and London: Routledge, 1999), xxi.

46. Elliott, *A Philosophical Disease*, xxiii.

47. Elliott, *A Philosophical Disease*, 11.

48. Catherine Waldby and Robert Mitchell, *Tissue Economies: Blood, Organs, and Cell Lines in Late Capitalism* (Durham: Duke University Press, 2006), 2, 188.

49. Waldby and Mitchell, *Tissue Economies, 41.*

50. Waldby and Mitchell, *Tissue Economies, 182.*

51. Margrit Shildrick, "Beyond the Body of Bioethics: Challenging the Conventions," in *Ethics of the Body: Postconventional Challenges*, eds. Margrit Shildrick and Roxanne Mykytiuk (Cambridge and London: MIT Press, 2005), 9.

52. Shildrick, "Beyond the Body of Bioethics," 10.

53. Margrit Shildrick, *Leaky Bodies and Boundaries: Feminism, Postmodernism and (Bio)ethics* (London and New York: Routledge, 1997), 10.

54. Shildrick, *Leaky Bodies and Boundaries*, 214.

55. Shildrick, *Leaky Bodies and Boundaries*, 212.

56. Gilles Deleuze, *Spinoza: Practical Philosophy*, trans. Robert Hurley (San Francisco: City Lights Books, 1988), 27.

57. Deleuze, *Spinoza*, 23. Deleuze explains: "The good is when a body directly compounds its relation with ours, and, with all or part of its power, increases ours," 22.

58. See Adrian Mackenzie, *Transductions: Bodies and Machines at Speed* (Continuum: London and New York, 2002), 175.

59. Claire Colebrook, *Gilles Deleuze* (London and New York: Routledge, 2002), 55.

60. Colebrook, *Gilles Deleuze*, 55.

61. Colebrook, *Gilles Deleuze*, 96.

62. Adrian Mackenzie, " 'God Has No Allergies': Immanent Ethics and the Simulacra of the Immune System," *Postmodern Culture*, vol. 6, no. 2 (January, 1996), available online, nonpag.

63. Eugene Thacker, *Biomedia* (Minneapolis and London: University of Minnesota Press, 2004), 184–5, 184.

64. Thacker, *Biomedia*, 184–5, 192–3. Thacker turns to Humberto Maturana's work on metadesign to elaborate this latter point.

65. Thacker, *Biomedia*, 188.

66. Rosi Braidotti, *Transpositions: On Nomadic Ethics* (Cambridge: Polity Press, 2006), 3–4.

67. Braidotti, *Transpositions*, 33.

68. Braidotti, *Transpositions*, 9.

69. Chapter 3 considers in more detail Giorgio Agamben's engagement with the Greek term *zoē*, which stands for bare life or the sheer fact of living. It also looks at some possible limitations of Agamben's reading of *zoē* in terms of a border concept that functions as an "inclusive exclusion" for future-oriented liberatory politics.

70. Daniel W. Smith, "Deleuze and Derrida, Immanence and Transcendence: Two Directions in Recent French Thought," in *Between Deleuze and Derrida*, eds. Paul Patton and John Protevi (London and New York: Continuum, 2003), 61.

71. For Deleuze and Guattari becoming is "the action by which something or someone continues to become other (while continuing to be what it is)," cited in Paul Patton and John Protevi, "Introduction," *Between Deleuze and Derrida*, 22. As we can see, difference and differentiation are hence seen as immanent, that is, they come from within Being and beings.

72. Patton and Protevi, "Introduction," 17.

73. Judith Butler, *Undoing Gender* (New York and London: Routledge, 2004), 198.

74. Rosalyn Diprose, *Corporeal Generosity: On Giving with Nietzsche, Merleau-Ponty, and Levinas* (Albany: SUNY Press, 2002), 141.

75. Rosalyn Diprose, *The Bodies of Women: Ethics, Embodiment and Sexual Difference* (London and New York: London, 1994), 18.

76. Diprose, *The Bodies of Women*, 18.

77. Chapter 2 presents a more sustained critique of Levinas's humanism, but it also considers its usefulness for thinking bioethics otherwise—with some help from Derrida and Stiegler.

Chapter 2

An abbreviated version of this chapter will be published as "Is There Life in Cybernetics? Designing a Posthumanist Bioethics," in Rosi Braidotti, Claire Colebook, and Patrick Hanafin (eds.) *Law after Deleuze* (Basingstoke: Palgrave Macmillan, 2009).

1. Helga Kuhse and Peter Singer, "Introduction," in *Bioethics: An Anthology*, eds. Helga Kuhse and Peter Singer (Oxford: Blackwell, 1999), 1.

2. Simon Glendinning, *In the Name of Phenomenology* (London and New York: Routledge, 2007), 184.

3. Rosi Braidotti, *Transpositions: On Nomadic Ethics* (Cambridge: Polity Press, 2006), 11.

4. See Warren Thomas Reich, "The Word 'Bioethics': Its Birth and the Legacies of Those Who Shaped It," *Kennedy Institute of Ethics Journal*, vol. 4, no. 4 (1994): 319–35, especially 325–6. According to Reich's genealogical study, "it is quite possible that Potter's word 'bioethics' influenced the development and use of the term at Georgetown," 327.

5. Warren Thomas Reich, "The Word 'Bioethics': The Struggle over Its Earliest Meanings," *Kennedy Institute of Ethics Journal*, vol. 5, no. 1 (1995): 19–34, 20–1.

6. Reich, "The Word 'Bioethics': Its Birth," 320.

7. See Adam Muller and Paisley Livingston, "Realism/Anti-Realism: A Debate," *Cultural Critique*, no. 30, *The Politics of Systems and Environments*, part I (Spring 1995): 15–32.

8. Kathleen Woodward, "Cybernetic Modeling in Recent American Writing: A Critique," *North Dakota Quarterly*, vol. 51, no. 1 (Winter 1983): 57–73, 69.

9. Donna J. Haraway, *Simians, Cyborgs and Women: The Reinvention of Nature* (London: Free Association Books, 1991).

10. Woodward, "Cybernetic Modeling in Recent American Writing," 69.

11. N. Katherine Hayles, *How We Became Posthuman* (Chicago and London: University of Chicago Press, 1999), 8.

12. N. Katherine Hayles, "Making the Cut: The Interplay of Narrative and System, or What Systems Theory Can't See," *Cultural Critique*, no. 30, *The Politics of Systems and Environments*, part I (Spring 1995): 71–100, 82.

13. Hayles, *How We Became Posthuman*, 10–11.

14. Cary Wolfe, "In Search of Post-Humanist Theory: The Second-Order Cybernetics of Maturana and Varela," *Cultural Critique*, no. 30, *The Politics of Systems and Environments*, part I (Spring 1995): 33–70, 47.

15. Quoted in Wolfe, "In Search of Post-Humanist Theory," 52.

16. Hayles, *How We Became Posthuman*, 136.

17. Wolfe, "In Search of Post-Humanist Theory," 60.

18. See Jacques Derrida, *Limited Inc.* (Evanston: Northwestern University Press, 1988) and Judith Butler, *Excitable Speech: A Politics of the Performative* (New York and London: Routledge, 1997).

19. William Rasch and Cary Wolfe, "Introduction: The Politics of Systems and Environments," *Cultural Critique*, no. 30, *The Politics of Systems and Environments*, part I (Spring 1995): 5–13, 9.

20. This idea is also drawn on in one of the more recent best-sellers of "alternative political theory": *Empire* by Michael Hardt and Antonio Negri (Cambridge: Harvard University Press, 2000). The authors quote Niklas Luhmann's systems theory as one of their theoretical influences, 13.

21. Rasch and Wolfe, "Introduction," 11.

22. Hayles, *How We Became Posthuman*, 100.

23. In her proposal for Deleuze-inspired nomadic ethics, Rosi Braidotti provides an alternative reading of such "entropic" practices. She says: "By de-pathologizing these allegedly 'extreme' clinical cases, we can approach them not so much as indicators of disorder, but as markers of a standard condition, namely the human subjects' enfleshed exposure to the irrepressible and at times hurtful vitality of life (*zoe*) and hence also the familiarity with or proximity to the crack, the line of unsustainability," *Transpositions*, 209–10. For a more "redeeming" reading of what are typically seen as addictive and self-destructive behaviors, see the whole of chapter 5, "Transcendence: Transposing Death" in Braidotti's *Transpositions*.

24. Hayles, "Making the Cut," 80.

25. Tiziana Terranova, *Network Culture: Politics for the Information Age* (London: Pluto Press, 2002), 104.

26. Hayles suggests that "narrative renders the closures systems theory would perform contingent rather than inevitable, thus mitigating the coercive effects that systems theory can sometimes generate," "Making the Cut," 98.

27. Hayles, "Making the Cut," 72.

28. See Bernard Stiegler, *Technics and Time, 1: The Fault of Epimetheus*, trans. Richard Beardsworth and George Collins (Stanford: Stanford University Press, 1998), 66–67.

29. See Jacques Derrida, "'There is No *One* Narcissism' (Autobiophotographies)," in Jacques Derrida, *Points . . . Interview, 1974–1994*, ed. Elisabeth Weber (Stanford: Stanford University Press, 1995), 212.

30. From a cybernetic perspective, any living system—human, animal, cell—can be conceptualized as a unit of interactions.

31. Neil Badmington argues that even cultural studies, a discipline that has sought to break down a series of oppressive barriers, has also "systematically reaffirmed the hierarchical border between the human and the inhuman" (262). Positioning man as the dominant producer of what counts as "culture" and "achievement," cultural studies has therefore perpetuated speciesism as the belief in the superiority and absolute distinctness of the human in relation to other species and life forms. Badmington goes on to propose the discipline of "posthumanities" as "the mark of a critical and gradual engagement with the relationship between the humanities and the figure of 'Man'" (266), in recognition that culture does not begin and end with what we call "human." Neil Badmington, "Cultural Studies and the Posthumanities," in *New Cultural Studies: Adventures*

in Theory, eds. Gary Hall and Clare Birchall (Edinburgh: Edinburgh University Press, 2006).

32. Stiegler, *Technics and Time*, 113.

33. Levinas writes: "A being independent of and yet at the same time exposed to the other is a temporal being: to the inevitable violence of death it opposes its time, which is postponement itself. . . . To be temporal is to be both for death and to still have time, to be against death." Emmanuel Levinas, *Totality and Infinity: An Essay on Exteriority*, trans. Alphonso Lingis (Pittsburgh: Duquesne University Press), 224, 235.

34. Stiegler, *Technics and Time*, 128.

35. Wolfe, "In Search of Post-Humanist Theory," 35.

36. Van Rensselaer Potter, "Bioethics, the Science of Survival," *Perspectives in Biology and Medicine*, vol. 14 (1970): 127–53. This article was later incorporated into Potter's book, *Bioethics: Bridge to the Future* (Englewood Cliffs, NJ: Prentice-Hall, Inc., 1971), as chapter 1.

37. Potter, "Bioethics, the Science of Survival," 27.

38. Reich, "The Word 'Bioethics': The Struggle," 29.

39. This is how Peter J. Whitehouse explains the relative lack of impact of Potter's theory on dominant bioethics: "Several attempts have been made to understand why Potter's conception of bioethics to this point has not influenced the development of mainstream biomedical ethics to what some of us would consider an appropriate degree. The identification of the word 'bioethics' with the Kennedy Institute closer to the heart of power in Washington, D.C., allowed it greater influence in clinical medicine. In the 1970s people were concerned about the implications of medical technology, particularly reproductive technology. This focus on the ethical implications of medical 'breakthroughs' on human values continues to dominate much of biomedical ethics. The tragic lack of concern in our healthcare 'systems' for environmental and public health could likely be associated with the same social forces that led to ignoring Potter's formulation of bioethics, such as short-term, profit-oriented, high-tech, genetically based acute medicine. Ironically, Potter has received much more recognition outside the United States as a major force in bioethics, perhaps because of the greater wisdom in other countries about long-term health issues." Peter J. Whitehouse, "Van Rensselaer Potter: An Intellectual Memoir," *Cambridge Quarterly of Healthcare Ethics* 11 (2002): 331–34, 333.

40. Potter, *Bioethics: Bridge to the Future*, 128.

41. Potter, *Bioethics: Bridge to the Future*, 131.

42. Potter, *Bioethics: Bridge to the Future*, 36.

43. Potter, *Bioethics: Bridge to the Future*, 36.

44. Quoted in Woodward, "Cybernetic Modeling," 60.

45. Woodward, "Cybernetic Modeling," 65.

46. Woodward, "Cybernetic Modeling," 66.

47. Woodward, "Cybernetic Modeling," 67.

48. Potter, *Bioethics: Bridge to the Future*, 52.

49. Potter, *Bioethics: Bridge to the Future*, 80–81.

50. Potter, *Bioethics: Bridge to the Future*, 90.

51. Potter, *Bioethics: Bridge to the Future*, 37.

52. Potter, *Bioethics: Bridge to the Future*, 82.

53. Hayles, *How We Became Posthuman*, 147.

54. See Potter, *Bioethics: Bridge to the Future*, 36.

55. Potter, *Bioethics: Bridge to the Future*, 149.

56. Wolfe, "In Search of Post-Humanist Theory," 61.

57. Wolfe, "In Search of Post-Humanist Theory," 63.

58. Francisco J. Varela's own ethical position, outlined in his book, *Ethical Know-How: Action, Wisdom, and Cognition* (Stanford: Stanford University Press, 1992), can be subject to similar criticism. He defines ethics as "immediate coping" in a given situation, a spontaneous reaction to the goal that presents itself before me, rather than as a deliberate, willed action of a central I which would be rooted in rationality and moral judgment (4–5). This binary opposition between spontaneous action and reason organizes the whole of his ethical framework. While interesting in its attempt to take into account the not always conscious processes of embodied cognition that affect the outcome of what moral philosophy defines as ethical dilemmas, Varela's ethics shares with Potter's bioethics the kind of solipsism that results from a clear commitment to the maintenance of the system's working and to the removal of disturbance and alterity from its operations. He writes, "the key to autonomy is that a living system finds its way into the next movement by acting appropriately out of its own resources. And it is the breakdowns, the hinges that articulate microworlds, that are the source of the autonomous and creative side of living cognition. . . . it is *during breakdowns* that the concrete is born" (11). Alterity is reduced here to a breakdown within the system and the possibility for the combinatoric working out of a new way of being, but there is something rather closed about this openness, as the resources the I draws on are already there within the system. Varela's turn to Buddhism and its practice of recognizing the emptiness of the self allows him to posit that we "are already other-directed even at our most negative, and we already feel warmth toward some people, such as family and friends" (67), a reading that seems oblivious to the differing cultural inscription of this "turn towards otherness." It also forecloses on the possibility of seeing violence, antagonism, and the potential break within the system as conditions of an ethical event, not as impediments to it.

59. Wolfe, "In Search of Post-Humanist Theory," 66.

60. Jacques Derrida, "Hospitality, Justice and Responsibility: A Dialogue with Jacques Derrida," in *Questioning Ethics: Contemporary Debates in Philosophy*,

eds. Richard Kearney and Mark Dooley (London and New York: Routledge, 1999), 66.

61. Potter favors the scientific–philosophic concept of progress for his bioethics, whereby "progress is in part definable as change that permits survival in a changing environment," *Bioethics: Bridge to the Future*, 47. He believes that "the limits of knowledge are infinite, . . . that no individual can begin to encompass even the knowledge that exists today . . . [and] that knowledge should be as widely disseminated as possible" (49). He also claims that the scientific–philosophical concept of progress is "completely foreign to communist ideology, and it is difficult to see how it could arise spontaneously in the communist system which places its emphasis on society and not on individuals. It is also opposed to then outworn political philosophies that regard individuals as taking supreme priority over society" (49).

62. Wolfe, *Animal Rites*, 88.

63. For an earlier elaboration of a nonsystemic theory of ethics for humans, animals, and machines, see my chapter "*Bio*-ethics and Cyberfeminism" in Joanna Zylinska, *The Ethics of Cultural Studies* (London and New York: Continuum, 2005), 138–58.

64. Jacques Derrida, *Of Grammatology*, trans. Gayatri Chakravorty Spivak (Baltimore and London: The Johns Hopkins University Press, 1976), 9.

65. Emmanuel Levinas and Richard Kearney, "Dialogue with Emmanuel Levinas," in *Face to Face with Levinas*, ed. Richard A. Cohen (Albany: SUNY Press, 1986), 15.

66. Emmanuel Levinas, *Totality and Infinity: An Essay on Exteriority*, trans. Alphonso Lingis (Pittsburgh: Duquesne University Press, 1969), 46.

67. Levinas, *Totality and Infinity*, 198.

68. Levinas, *Totality and Infinity*, 198, emphasis added.

69. Levinas, *Totality and Infinity*, 198.

70. Levinas, *Totality and Infinity*, 203–4. The introductory material on Levinas included here has been reworked from the chapter "A User's Guide to Culture, Ethics and Politics" included in my book *The Ethics of Cultural Studies* (London and New York: Continuum, 2005).

71. John Llewelyn, "Am I Obsessed by Bobby? (Humanism and the Other Animal)," in *Re-reading Levinas*, eds. Robert Bernasconi and Simon Critchley (Bloomington and Indianapolis: Indiana University Press, 1991), 237.

72. Emmanuel Levinas, "The Name of a Dog, or Natural Rights," in *Difficult Freedom: Essays on* Judaism, trans. Sean Hand (London: The Athlone Press, 1990), 153.

73. Levinas, "The Name of a Dog, or Natural Rights," 152.

74. In a response to the historians Anne McClintock and Rob Nixon, who accused him of "textualizing" and homogenizing the experience of apartheid, Derrida writes: "[*T*]*ext*, as I use the word, is not the book. No more than writing

or trace, it is not limited to the *paper* which you cover with your graphism. It is precisely for strategic reasons . . . that I found it necessary to recast the concept of text by generalizing it almost without limit, in any case without present or perceptible limit, without any limit that *is*. That's why there is nothing 'beyond the text.' That's why South Africa and *apartheid* are, like you and me, part of this general text. . . . [T]he text is always a field of forces: heterogeneous, differential, open, and so on." Jacques Derrida, "But, Beyond . . ." (Open Letter to Anne McClintock and Rob Nixon), *Critical Inquiry*, vol. 13, no. 1 (Autumn 1986): 155–70, 167–8.

75. Emmanuel Levinas, *Otherwise Than Being: Or Beyond Essence*, trans. Alphonso Lingis (Pittsburgh: Duquesne University Press, 1998), 133–4.

76. Levinas, *Otherwise Than Being*, 134.

77. Levinas, *Totality and Infinity*, 23.

78. Levinas, *Totality and Infinity*, 40.

79. Stelarc, "Prosthetic Head: Intelligence, Awareness and Agency," *CTheory* (2005), nonpag.

80. Levinas, *Otherwise Than Being*, 135.

81. Bernard Stiegler, "Technics of Decision: An Interview with Peter Hallward," trans. Sean Gaston. *Angelaki*, vol. 8, no. 2 (2003): 151–68. 158.

82. See Mark Poster, *Information Please: Culture and Politics in the Age of Digital Machines* (Durham and London: Duke University Press, 2006), 142–3.

83. Poster, *Information Please*, 139, 142.

84. Glendinning, *In the Name of Phenomenology*, 205.

85. For Levinas, "I am put in the passivity of an undeclinable assignation, in the accusative, a self. Not as a particular case of the universal, an ego belonging to the concept of ego, but as I, said in the first person—I, unique in my genus," *Otherwise Than Being*, 139.

86. See Stiegler, "Technics of Decision," 158.

Chapter 3

1. Helga Kuhse and Peter Singer, "Introduction," in *Bioethics: An Anthology*, eds. Helga Kuhse and Peter Singer (Oxford: Blackwell, 1999), 1.

2. As an example of this moralism, we can reference media discourses and strategies which are used to establish a connection between physical and moral health, for example, between obesity and laziness in children or alcohol intake and "yobbishness." For an explication of the distinction between ethics, morality, and moralism, see Joanna Zylinska, "Cultural Studies and Ethics," in *New Cultural Studies: Adventures in Theory*, eds. Gary Hall and Clare Birchall (Edinburgh: Edinburgh University Press, 2006).

3. Michel Foucault, *The History of Sexuality: An Introduction*, trans. Robert Hurley (Harmondsworth: Penguin Books, 1984), 136.

4. Michel Foucault, *"Society Must Be Defended": Lectures at the Collège de France, 1975–76*, trans. David Macey (London: Allen Lane, 2003), 241.

5. Foucault, *The History of Sexuality*, 139.

6. Foucault, *"Society Must Be Defended,"* 242–3.

7. Foucault, *"Society Must Be Defended,"* 245.

8. Giorgio Agamben, *Homo Sacer: Sovereign Power and Bare Life*, trans. Daniel Heller-Roazen (Stanford: Stanford University Press, 1998), 1.

9. Thomas Carl Wall, "Bare Sovereignty: *Homo Sacer* and the Insistence of Law," in *Politics, Metaphysics and Death: Essays on Giorgio Agamben's* Homo Sacer, ed. Andrew Norris (Durham and London: Duke University Press, 2005), 44.

10. Agamben, *Homo Sacer*, 3.

11. See Agamben, *Homo Sacer*, 4.

12. Agamben, *Homo Sacer*, 7.

13. See Agamben, *Homo Sacer*, 9.

14. Agamben, *Homo Sacer*, 9.

15. See Agamben, *Homo Sacer*, 20, and Giorgio Agamben, *State of Exception*, trans. Kevin Attell (Chicago: University of Chicago Press, 2005): 1–7. Even though "the *raison d'etre* of contemporary political power is . . . a total politicization of biological life that undercuts the distinction between *bios* and *zoē*," Catherine Mills argues that for Agamben "modern democracy has consistently failed in the endeavor to reconcile *bios* and *zoē*," a failure that positions bare life as precisely an exception, or something that is included only through an exclusion. Catherine Mills, "Agamben's Messianic Politics: Biopolitics, Abandonment and Happy Life," *Contretemps*, vol. 5 (December 2004): 42–62, 47.

16. Giorgio Agamben, "No to Biopolitical Tattooing," http://www.truthout.org/cgi-bin/artman/exec/view.cgi/4/3249 (2004): nonpag.

17. Agamben, *State of Exception*, 39.

18. Agamben, *State of Exception*, 3–4.

19. See Agamben, *State of Exception*, 1.

20. Eugene Thacker, "Nomos, Nosos and *Bios*," *Culture Machine*, vol. 7 (2005): nonpag.

21. Alison Ross, "Introduction," *The South Atlantic Quarterly*, vol. 107, no. 1 (Winter 2007): 1–13, 7.

22. Ibid., 7.

23. Rosi Braidotti, *Transpositions: On Nomadic Ethics* (Cambridge: Polity Press, 2006), 39.

24. Ewa Plonowska Ziarek, "Bare Life on Strike: Notes on the Biopolitics of Race and Gender," *The South Atlantic Quarterly*, vol. 107, no. 1 (Winter 2007): 89–105, 90.

25. Penelope Deutscher explains that, with the exception of Karen Quinlan's comatose body that is analyzed in his *Homo Sacer*, there is a marked absence of women's bodies as well as reproductive bodies from Agamben's writings. Indeed, Agamben's work arguably dissociates "life" from "woman's reproductivity," although Deutscher recognizes in it a potential to unsettle and rethink the terms such as *life*, *bare life*, *threshold*, and *biopolitics*. In Penelope Deutscher, "The Inversion of Exceptionality: Foucault, Agamben, and 'Reproductive Rights,'" *The South Atlantic Quarterly*, vol. 107, no. 1 (Winter 2007): 55–70, 59. Claire Colebrook develops this point further by arguing that Agamben's theory of potentiality and generativity resonates with the highly gendered and theological humanism, at the expense of more materialist and embodied conceptions of productivity. What is repressed in his work, according to Colebrook, is what has always been associated with woman's productivity: "a birth, production, or creation that is neither expressive, free, nor open, a production all too weighted down by the already actualized. . . . So while Agamben criticizes the modern notion of life as simply willing to maintain itself as it is, he draws on a supposedly lost higher sense of life as that which creates from itself in order to be more than itself: divine, poetic life—the life of man." In Claire Colebrook, "Agamben, Potentiality, and Life," *The South Atlantic Quarterly*, vol. 107, no. 1 (Winter 2007): 107–20, 110.

26. Ziarek, "Bare Life on Strike," 96.

27. Ziarek, "Bare Life on Strike," 97.

28. Braidotti, *Transpositions*, 40.

29. Braidotti, *Transpositions*, 39.

30. See Paul Rabinow, "Introduction" to Michel Foucault, *Ethics: Subjectivity and Truth*, ed. Paul Rabinow (London: Allen Lane, 1997), xxv–xxvii.

31. Michel Foucault, *Ethics: Subjectivity and Truth*, ed. Paul Rabinow (London: Allen Lane, 1997), 68.

32. Michel Foucault, *The Hermeneutics of the Subject: Lectures at the Collège de France 1981–82*, eds. Frederic Gros and Francois Ewald (Palgrave Macmillan: Basingstoke, 2005), 36.

33. Foucault, *The Hermeneutics of the Subject*, 8.

34. See Foucault, *Ethics*, 269–71.

35. Foucault, *The Hermeneutics of the Subject*, 84.

36. Foucault, *Ethics*, 132.

37. Foucault, *Ethics*, 131.

38. Foucault, *Ethics*, 163.

39. See Timothy O'Leary, *Foucault and the Art of Ethics* (London and New York: Continuum, 2002), 128–9.

40. Ibid., 131.

41. Maurizio Lazzarato, "From Biopower to Biopolitics," trans. Ivan A. Ramirez. Available at http://www.geocities.com/immateriallabour/lazzarato-from-biopower-to-biopolitics.html (nondat.). Accessed May 31, 2007.

42. Lazzarato, "From Biopower to Biopolitics."

43. Foucault, *Ethics*, 225.

44. Foucault, *Ethics*, 232.

45. In *Foucault and the Art of Ethics* Timothy O'Leary admits that Foucault tends to apply a modern aesthetic sensibility to the ancient texts, which leads Foucault to position the Greeks and Romans as "an early chapter in the history of the 'arts of the self' " and as precursors to Oscar Wilde (London and New York: Continuum, 2002), 51. However, O'Leary defends Foucault's (not entirely unacknowledged) ahistoricism, whereby the ancients are seen as individuals engaged in the explicit work of self-creation, and argues that the French thinker was principally motivated by present concerns rather than by fidelity to historical truth. The Greeks and Romans were therefore for him an inspiration for developing "a contemporary post-Christian ethics of self-transformation," one that would not be subject "to the harsh demands of a punitive moral code," 81–83.

46. As Foucault puts it, "Care of the self becomes coextensive with life," *The Hermeneutics of the Subject*, 86.

47. Foucault, *Ethics*, 260.

48. Foucault, *Ethics*, 235.

49. For an example of such an "applied bioethics" project which engages with the work of Foucault but which is situated within the more established "medical humanities" framework, see the special issue of the *Journal of Medical Humanities* titled "Bioethics and the Later Foucault," edited by Arthur W. Frank and Therese Jones, vol. 24, nos. 3–4 (Winter 2003): 179–86.

50. Foucault, *Ethics*, 97.

51. Foucault, *The Hermeneutics of the Subject*, 125–6.

52. Foucault, *The Hermeneutics of the Subject*, 107–8.

53. Other examples of popular social networking Web sites include Facebook, Opendiary.com and MyDearDiary.com, Blurty, Xanga, DeadJournal, Blogger, and DiaryLand. Perhaps appropriately, I used Wikipedia (http://www.wikipedia.org), a free online encyclopedia edited (and constantly re-edited) by users themselves but also a form of online community in itself, to develop some of the definitions presented in the main body of the chapter.

54. Adam Reed, " 'My Blog Is Me': Texts and Persons in U.K. Online Journal Culture (and Anthropology)," *Ethnos*, vol. 70, no. 2 (June 2005): 220–42, 227.

55. N. Katherine Hayles, *Writing Machines* (Cambridge and London: MIT Press, 2002), 25–33.

56. Mark Poster, *Information Please: Culture and Politics in the Age of Digital Machines* (Durham and London: Duke University Press, 2006), 35.

57. See Gary Hall, "The Singularity of New Media" in his *Digitize This Book!: The Politics of New Media, or Why We Need Open Access Now* (Minneapolis: University of Minnesota Press, 2008), 208–15.

58. Bebo is another social networking Web site and an online community, which allows its members to post photos and videos and to blog and communicate with one another. As Oliver Burkeman writes in *The Guardian*, "Bebo, like MySpace, provides its more than 23m users with a kind of prosthetic personality extension: a profile page where they can post diary entries, photographs, music, homemade video and answers to questionnaires—about what scares them or makes them most happy." Box insert into an article by John Lanchester, "A Bigger Bang," *The Guardian Weekend*, November 4 (2006), 17–36, 21.

59. José Van Dijck writes: "Yet if we closely look at how paper diaries were used in the past, the characteristics of uniformity, privacy and single authorship are, to say the least, disputable; it is surprising to find, though, how these accepted notions about diaries still affect today's theorisation of weblogs. Over the past centuries, the diary as a cultural form has been anything but homogeneous. The genre has been defined as therapy or self-help, as a means of confession, as a chronicle of adventurous journeys (both spiritual and physical), or as a scrapbook for creative endeavours." José Van Dijck, "Composing the self: Of diaries and lifelogs," *Fibreculture Journal*, no. 3 (2004): nonpag. Van Dijck goes on to argue that the perception of a diary as a private genre, strictly written for oneself, has been as misleading as it has been persistent. She also points out that since its very inception, the genre has been dialogic, hence obliterating the line between private and public (cf. diaries shared within religious communities).

60. See Hall, *Digitize This Book!*, 114–15.

61. Christopher Lasch, *The Culture of Narcissism: American Life in an Age of Diminishing Expectations* (London: Norton, 1979).

62. Kris R. Cohen, "A Welcome for Blogs," *Continuum: Journal for Media & Culture Studies*, vol. 20, no. 2 (June 2006): 161–73, 164.

63. Cohen, "A Welcome for Blogs," 164.

64. The del.icio.us Web site—originally independent but now owned by Yahoo—is one example of such "sharing sites" on which users can store, exchange, and discover one another's Web bookmarks. Academic blogs, in turn, have been described as "experiments in digital scholarship," or a testing ground for research in progress (see Craig Saper, "Blogademia," *Reconstruction*, vol. 6, no. 4 (2006): nonpag.). Melissa Gregg's 2006 article, "Feeling Ordinary: Blogging as Conversational Scholarship," provides an excellent justification for the use of blogging as a form of scholarly exchange and an alternative scholarly practice, in *Continuum: Journal for Media & Cultural Studies*, vol. 20, no. 2 (June): 147–60.

65. As Mikkel Borch-Jacobsen puts it, "delirium always pivots on what one would (ideally) like to be, and not on what one would like to *have* (in the name

of sexual pleasure).... What I myself am not (namely, a *subject*: free, autonomous, independent, and so on) is always another (another *subject*: rich, famous, admired, recognized) who is taking my place—that place or social position where I would like to be. The problem, of course, is that I will then compete savagely with that other for that 'being,' for that 'place,' because if the other is in my place, it goes without saying that I shall never cease my efforts to dislodge him, in order finally to be myself. What begins in admiration ends in murder." Mikkel Borch-Jacobsen, *Lacan: The Absolute Master*, trans. Douglas Brick (Stanford: Stanford University Press, 1991), 24. The blogosphere, it seems, can also be a crime scene.

66. Jacques Derrida, "'There is No *One* Narcissism' (Autobiophotographies)," in Jacques Derrida, *Points... Interview, 1974–1994*, ed. Elisabeth Weber (Stanford: Stanford University Press, 1995), 199.

67. Derrida, "'There is No *One* Narcissism,'" 199.

68. Van Dijck seems to understand blogging principally as a means of self-expression. She suggests that "For the contemporary blogger, the internet is just one of a host of media through which to *express agency*, and blogging is one of many competing practices, such as speaking (both face-to-face and phone conversations), writing (letters, sms, e-mail), watching (television, film, photos) and listening (music, talk)," although she also recognizes that through their *LiveJournals* or *Xangas*, "teenagers not only *express themselves*, but create a communal sense of values and thoughts deemed worthy of being shared," "Composing the Self," nonpag., emphasis added.

69. Ewa Plonowska Ziarek, *An Ethics of Dissensus: Postmodernity, Feminism, and the Politics of Radical Democracy* (Stanford: Stanford University Press, 2001), 39.

70. Ziarek, *An Ethics of Dissensus*, 41.

71. Ziarek explains: "Because it bears a relation to the outside, which is the experience of the limit of social regulation and thus of the historical constitution of the subject, the experimental praxis cannot be simply reduced to a goal-oriented activity. On the contrary, the outcome of such an experimental praxis aiming to surpass the historical limits of bodies, language and sexuality cannot be predicted in advance because it opens up a relation to a future that can [no] longer be thought on the basis of the present.... Foucault's invention of the improbable stresses the radical futural dimension of praxis beyond the anticipation of the subject." Ziarek, *An Ethics of Dissensus*, 41.

72. Craig Saper, "Blogademia," nonpag., emphasis added.

73. Foucault, *Ethics*, 211.

74. Foucault, *Ethics*, 213.

75. Foucault, *The Hermeneutics of the Subject*, 113.

76. Foucault, *The Hermeneutics of the Subject*, 120.

77. Van Dijck, "Composing the Self," nonpag.

78. The American Internet entrepreneur Alan Levy is rather explicit about the status quo: "The blogosphere is a world of haves and have-nots. Frankly, about two bloggers out of every million have meaningful traffic and readership." Christian Hall, "An Interview with Alan Levy," *net*, issue 157 (December 2006): 34–36, 24.

79. Foucault, *Ethics*, 272.

80. Poster, *Information Please*, 38.

81. Poster, *Information Please*, 36.

82. I am indebted to my PhD student Federica Frabetti at Goldsmiths for her discussion of the relationship between memory and technology in Stiegler.

83. Bernard Stiegler, "Technics of Decision: An Interview with Peter Hallward," trans. Sean Gaston, *Angelaki*, vol. 8, no. 2 (2003): 151–68, 161. Existence, or life, for humans means the use of language, i.e., writing, which in *Of Grammatology* Derrida interprets as the leaving of traces or marks.

84. I found out about MyDeathSpace.com from Cortney Heimerl's presentation at the "Homelands: Spaces of Inclusion/Exclusion, Spaces of Power/Security" graduate conference at NYU Steinhardt on November 10, 2006.

85. Poster, *Information Please*, 129.

86. Poster, *Information Please*, 129.

87. Lanchester, "A Bigger Bang," 31.

88. Foucault, *Ethics*, 319.

89. Ziarek, *An Ethics of Dissensus*, 6.

90. See, for example, Alain Badiou, *Ethics: An Essay on the Understanding of Evil*, trans. Peter Hallward (London and New York: Verso, 2001).

91. Foucault, *The Hermeneutics of the Subject*, 12.

Chapter 4

An earlier version of this chapter was published as "Of Swans and Ugly Ducklings: Bioethics between Humans, Animals, and Machines" in *Reality Made Over: The Culture of Reality Makeover Shows*, guest ed. Bernadette Wegenstein, *Configurations*, vol. 15 no. 2 (December 2008).

1. *The Swan* voiceover, courtesy of Reality News Online, http://www.realitynewsonline.com. Accessed July 5, 2005.

2. Some other makeover shows which have gained international popularity include Bravo's *Queer Eye for the Straight Guy*, BBC's *What Not to Wear*, ABC's *Extreme Makeover: Home Edition*, BBC's *Ground Force*, and Britain's Channel 4's *10 Years Younger*.

3. In *The History of Sexuality: An Introduction* Foucault associates biopolitics with the less disciplinary, more productive management of the life and health of

whole populations and contrasts it against "an anatomo-politics of the human body" (trans. Robert Hurley, Harmondsworth: Penguin Books, 1984, 139). Even though it is only the latter that is explicitly disciplinary for Foucault, I show in chapters 3 and 4 how this more positive, allegedly less constraining biopolitical management always inheres disciplinarity and violence, even if these mechanisms remain concealed to some extent.

4. Foucault, *The History of Sexuality*, 137.

5. Giorgio Agamben, *Homo Sacer: Sovereign Power and Bare Life*, trans. Daniel Heller-Roazen (Stanford: Stanford University Press, 1998), 20.

6. The inspiration for this chapter came from a piece by Myrna Blyth in National Review Online, "Skin Deep," which opened with the following words: "This past Monday night was a makeover marathon on TV. While President Bush outlined his plans for the makeover of Iraq to the Army War College on the Fox News network, much of America was tuned into The Swan, the mother of all makeover shows, on the other Fox channel," May 24, 2004, http://www.nationalreview.com/blyth/blyth200405270957.asp. Accessed June 1, 2005.

7. Sander L. Gilman, *Making the Body Beautiful: A Cultural History of Aesthetic Surgery* (Princeton: Princeton University Press, 2000), 19.

8. Gilman, *Making the Body Beautiful*, 18.

9. Gilman, *Making the Body Beautiful*, 21.

10. Gilman, *Making the Body Beautiful*, 30.

11. Agamben, *Homo Sacer*, 20.

12. Giorgio Agamben, *State of Exception*, trans. Kevin Attell (Chicago: University of Chicago Press, 2005), 3–4.

13. Andrew Norris, "The Exemplary Exception: Philosophical and Political Decision on Giorgio Agamben's *Homo Sacer*," in *Politics, Metaphysics and Death: Essays on Giorgio Agamben's* Homo Sacer, ed. Andrew Norris (Durham and London: Duke University Press, 2005), 264.

14. Agamben, *Homo Sacer*, 159.

15. Agamben, *Homo Sacer*, 73.

16. " 'To ban' someone is to say that anyone may harm him" (Cavalca, quoted in Agamben, *Homo* Sacer, 105). Catherine Mills interestingly explores the ambiguity of the position of *homo sacer* in Agamben's work by pointing out that *homo sacer* exists for him in the zone of disturbing proximity or even indistinction with the legitimate members of the polis. For Mills, "Agamben claims from this that under a regime of biopolitics all subjects are potentially *homo sacers*. That is, all subjects are at least potentially if not actually abandoned by the law and exposed to violence as a constitutive condition of political existence. As empirical evidence of this politicophilosophical claim, he cites the figure of *homo sacer*, genocidal violence, the apparently ever-expanding phenomenon of concentration camps—which he argues reveal the '*nomos* of the modern'—as well as the redefinition of life and death in the categories of the 'overcomatose' or brain

dead, and neo-morts. Agamben has been heavily criticized for his apparently eclectic collection of empirical evidence and the rendering of these examples as 'indistinguishable.' Yet, what unites the examples Agamben selects is the thesis on the generalization of the exception and the correlative indistinction of fact and norm in Western politics and philosophy." In "Agamben's Messianic Politics: Biopolitics, Abandonment and Happy Life," *Contretemps*, vol. 5 (December 2004): 42–62, 47.

17. Agamben, *Homo Sacer*, 105.

18. Michel Foucault, "Technologies of the Self," in *Technologies of the Self*, eds. Luther H. Martin et al. (London: Tavistock Books, 1990), 18.

19. Gilman, *Making the Body Beautiful*, 333.

20. Emmanuel Levinas, "Ethics as First Philosophy," in *The Levinas Reader*, ed. Sean Hand (Oxford: Blackwell, 1989), 83. In his writings Levinas distinguishes between *autrui* (the personal other, the you, translated as Other) and *autre* (other). He frequently alternates between the two within any one paragraph. Unless I am quoting directly from Levinas, I have opted for the use of the lower case "other" throughout this volume in order to lessen the impact of Levinas's implied humanism.

21. Emmanuel Levinas, *Totality and Infinity: An Essay on Exteriority*, trans. Alphonso Lingis (Pittsburgh: Duquesne University Press, 1969), 66.

22. See Levinas, *Totality and Infinity*, 194.

23. Even if the shift to the first-person singular may read somewhat awkwardly here, it is important to retain this pronoun—instead of using the more generic "one," "we," or "man"—in order to highlight the absolute singularity and irreplaceability of the obligation that structures Levinas's ethics. This obligation and responsibility is only ever *mine* rather than "human" or "ours."

24. In makeover programs, this is often a codeword for the "Jewish nose." See Gilman, *Making the Body Beautiful.*

25. Foucault's work on the "care of self," which I discuss in chapter 3, would provide one possible way of interpreting cosmetic surgery clients and makeover show participants as agents of their transformation, not just victims of ideology and overbearing hegemonic power relations. Body makeover could then be read as a practice of self-transformation, an attempt to establish an ethical relation to oneself through a set of selected techniques. Mark Poster analyzes *The Swan*'s participants through this prism of Foucaldian care of the self in "Swan's Way," *Configurations*, vol. 15 no. 2 (December 2008).

26. See Agamben, *Homo Sacer*, 186–7.

27. Giorgio Agamben, *The Open: Man and Animal*, trans. Kevin Attell (Stanford: Stanford University Press, 2004), 6.

28. Agamben, *The Open*, 26.

29. See Dinesh Wadiwel, "Animal by Any Other Name? Patterson and Agamben Discuss Animal (and Human) Life," *Borderlands: e-journal*, vol. 3, no. 1 (2004):

nonpag.; Bernard Stiegler, *Technics and Time, 1: The Fault of Epimetheus*, trans. Richard Beardsworth and George Collins (Stanford: Stanford University Press, 1998); Cary Wolfe, *Animal Rites: American Culture, the Discourse of Species, and Posthumanist Theory* (Chicago: University of Chicago Press, 2003), 75–6.

30. Donna J. Haraway, *Simians, Cyborgs and Women: The Reinvention of Nature* (London: Free Association Books, 1991). Originally published as: Donna Haraway, "A Manifesto for Cyborgs," *Socialist Review*, 80 (1985): 65–107.

31. Haraway, *Simians*, 152.

32. Quoted in Wolfe, *Animal Rites*, 5.

33. Wolfe, *Animal Rites*, 41.

34. Wolfe, *Animal Rites*, 101.

35. See Wolfe, *Animal Rites*, 105.

36. Wolfe, *Animal Rites*, 104–5.

37. In her second book, *Primate Visions: Gender, Race and Nature in the World of Modern Science* (New York and London: Routledge, 1989), Haraway explored the complex relations between humans, animals, and technoscience by focusing on primatology.

38. Donna Haraway, *The Companion Species Manifesto: Dogs, People, and Significant Otherness* (Chicago: Prickly Paradigm Press, 2003), 9–10.

39. See Haraway, *The Companion Species Manifesto*, 4.

40. Haraway, *The Companion Species Manifesto*, 5.

41. Haraway, *The Companion Species Manifesto*, 9, 3.

42. Haraway, *The Companion Species Manifesto*, 12.

43. Haraway, *The Companion Species Manifesto*, 2–3.

44. Haraway, *The Companion Species Manifesto*, 9.

45. Haraway, *The Companion Species Manifesto*, 25.

46. Haraway, *The Companion Species Manifesto*, 50.

47. Haraway, *The Companion Species Manifesto*, 53.

48. For an alternative nonsystemic theory of ethics for humans, animals, and machines, see my chapter "*Bio*-ethics and Cyberfeminism" in Joanna Zylinska, *The Ethics of Cultural Studies* (London and New York: Continuum, 2005), 138–58.

49. Haraway, *The Companion Species Manifesto*, 61.

50. The last few paragraphs come from my review of Haraway's *Companion Species Manifesto*, "Dogs R Us?," *parallax*, vol. 12, no. 1: 129–31.

51. http://www.swanuk.org.uk/index%20frame.htm, Swan Sanctuary's Web site. Accessed June 1, 2005.

52. Wolfe, *Animal Rites*, 26.

53. See Wolfe, *Animal Rites*, 26.

54. Wolfe, *Animal Rites*, 65.

55. Wolfe, *Animal Rites*, 16.

56. Jacques Derrida, "The Animal That Therefore I Am (More to Follow)," trans. David Wills, *Critical Inquiry*, vol. 28 (Winter 2002): 369–418, 402.

57. Derrida, "The Animal That Therefore I Am," 400.

58. The *Extra Ear: Ear on Arm* project by the international artist Stelarc, aimed at exploring alternate body architectures, involves constructing a replica of the artist's ear, equipping it with a microphone and implanting it on his arm. For information on the earlier stages of this project, see Stelarc's Web site: http://www.stelarc.va.com.au/quarterear/index.html. The last chapter of this book provides an interpretation of Stelarc's *Extra Ear* project.

Chapter 5

An earlier version of this chapter was published as "The Secret of Life: *Bio*ethics between Corporeal and Corporate Obligations," *Cultural Studies*, vol. 21, no. 1 (2007): 95–117.

1. Cultural studies is one discipline at the crossroads of the humanities and the social sciences that has taken the issue of responsibility for knowledge, and for its own relationship to knowledge, on board. Even though it initially defined itself in explicitly political terms, more recently discussions over cultural studies' political engagement have been linked to ethical considerations. For more on this see Mark Devenney and Joanna Zylinska (eds.), "Cultural Studies: Between Politics and Ethics" (special issue), *Strategies: Journal of Theory, Culture and Politics*, vol. 14, no. 2 (2001); Joanna Zylinska, *The Ethics of Cultural Studies* (London and New York: Continuum, 2005); and Joanna Zylinska, "Cultural Studies and Ethics," in *New Cultural Studies: Adventures in Theory*, eds. Gary Hall and Clare Birchall (Edinburgh: Edinburgh University Press, 2006).

2. The genetic code is a set of rules which maps DNA sequences to proteins in the living cell and is employed in the process of protein synthesis. Nearly all living organisms use the same genetic code. The mapping of the DNA structure by Watson and Crick in 1953 was the first step in the process of mapping the genetic code, which was finally accomplished in 1966.

3. As someone whose academic background lies in the humanities rather than the hard sciences, I remain open to learning from scientific debates—here in particular those taking place in molecular biology, biochemistry, genetics, and bioinformatics. I also share the commitment of a number of theorists working in science and technology studies, feminist studies of science, and now, increasingly, media and cultural studies, to an immersed, participatory, and "hospitable" approach to science.

4. Quoted in Evelyn Fox Keller, *Secrets of Life, Secrets of Death: Essays on Gender, Language and Science* (New York and London: Routledge, 1992), 96.

5. Genes are hypothetical particles carried on chromosomes and mediating inheritance. As Steve Rose explains, "Roughly speaking, a gene can be defined as a unit of information. That's how Mendel saw it. Biochemically, it's a segment of DNA, which produces an RNA message which is read off it. But now that we know more about the genome, the old central dogma, that genes make RNA protein, is by no means straightforward. For example, there are genes within genes, some genes overlap with each other, and there are great acres of DNA called 'junk DNA,' which doesn't seem to do much at all. So 'gene' is a convenient shorthand for what may turn out to be a rather complicated phenomenon." In Jeremy Stangroom, "Darwinism and Genes: In Conversation with Steve Jones," in *What Scientists Think* (London and New York: Routledge, 2005): 1–22, 10.

6. Keller, *Secrets of Life*, 42, see also 101–5.

7. James Watson, *DNA: The Secret of Life* (London: Arrow Books, 2004), 35.

8. See Lily E. Kay, *Who Wrote the Book of Life? A History of the Genetic Code* (Stanford: Stanford University Press, 2000), 59–72.

9. Kay, *Who Wrote the Book of Life?*, 5. This paradigm shift took place in 1953, with the mapping of the DNA structure by Watson and Crick.

10. The genome is the entire genetic setup of an organism (i.e., all of its genes). In other words, it is all of the genetic information or hereditary material possessed by an organism.

11. Kay, *Who Wrote the Book of Life?*, 90.

12. Watson, *DNA: The Secret of Life*, 53.

13. See Ruth Hubbard and Elijah Wald, *Exploding the Gene Myth* (Boston: Beacon Press, 1997), 69.

14. Kay, *Who Wrote the Book of Life?*, 2, 23.

15. Kay, *Who Wrote the Book of Life?*, 21, 100.

16. Kay, *Who Wrote the Book of Life?*, 127.

17. Nancy Gibbs, "The Secret of Life," *Time*, February 17 (2003): 42.

18. http://genome.pfizer.com/learn_more.cfm

19. Thomas Goetz, "Your DNA Decoded," *Wired* 15.2 (2007): 256–265, 283.

20. Keller, *Secrets of Life*, 96.

21. Watson, *DNA: The Secret of Life*, xx.

22. Dorothy Nelkin and Susan M. Lindee, *The DNA Mystique: The Gene as a Cultural Icon* (New York: W. H. Freeman and Company, 1995), 41.

23. See Henry Quastler, "The Status of Information Theory in Biology," in *Symposium on Information Theory in Biology*, ed. H. P. Yockey (New York: Pergamon Press, 1956), 399–401.

24. Western political sovereignty has always concerned itself with bare life, but this concern has taken a new form in the era of bio- and nanotechnologies, and

of the proliferation of the discourse (if not yet the material effects) of ubiquitous computing. For more on this point, see Eugene Thacker, "Nomos, Nosos and *Bios*," *Culture Machine*, vol. 7 (2005): nonpag.

25. See E. O. Wilson, *Sociobiology: The New Synthesis* (Cambridge, MA: Belknap Press, 1975).

26. Completed in 2003, the Human Genome Project was a thirteen-year effort coordinated by the U.S. Department of Energy and the National Institutes of Health. Its aims were to identify all of the approximately 20,000–25,000 genes in human DNA, determine the sequences of the three billion chemical base pairs that make up human DNA, store this information in databases, and transfer related technologies to the private sector. (Source: Web site of the Oak Ridge National Laboratory, working for the U.S. Department of Energy, http://www.ornl.gov/sci/techresources/Human_Genome/home.shtml.)

The Human Genome Diversity Project was announced in 1991. Its aim is to collect samples from 10% of the world's total groups, determined by their biological and geographical isolation, as well as their linguistic integrity. See Joanne Barker, "The Human Genome Diversity Project: 'Peoples,' 'Populations' and the Cultural Politics of Identification," *Cultural Studies*, vol. 18, no. 4 (2004): 571–606, 574–5.

27. Sarah Franklin explains in her entry on "life" in the *Encyclopedia of Bioethics*: "As the historian Michel Foucault points out, life itself did not exist before the end of the nineteenth century; it is a concept indebted to the rise of the modern biological sciences." Foucault's concepts of biopower and biopolitics, developed in relation to this idea of life itself, were a direct inspiration for Agamben's work on biopolitics. Sarah Franklin, "Life," in *The Encyclopedia of Bioethics*, ed. W. T. Reich, 5 vols. (New York: Macmillan Library Reference USA, 1995), 1347.

28. Giorgio Agamben, *Homo Sacer: Sovereign Power and Bare Life*, trans. Daniel Heller-Roazen (Stanford: Stanford University Press, 1998), 6, emphasis in original.

29. See Agamben, *Homo Sacer*, 26, 32, 35. Keller explores what we might describe as the unwelcome secret of those secret-cracking endeavors in science. She argues that the "perennial motif that underlies much of scientific creativity—namely, the urge to fathom the secrets of nature, and the collateral hope that, in fathoming the secrets of nature, we will fathom the ultimate secrets (and hence gain control) of our own mortality" has proceeded "on two fronts: the search for the wellspring of life and, simultaneously, for ever more effective instruments of death," Keller, *Secrets of Life*, 40.

30. Agamben, *Homo Sacer*, 53–5.

31. Agamben, *Homo Sacer*, 28.

32. Quoted in Barker, "The Human Genome Diversity Project," 574.

33. Rosalyn Diprose, "A 'Genethics' That Makes Sense," in *Biopolitics: A Feminist and Ecological Reader on Biotechnology*, eds. Vandana Shiva and Ingunn

Moser (London and Atlantic Highlands, NJ: Third World Network and Zed Books, 1995), 167.

34. C. Adami, "What is Complexity?," *BioEssays*, vol. 24, no. 12 (2002): 1085–94, 1086.

35. "Life has traditionally been seen as the secret of women, a secret *from* men. By virtue of their ability to bear children, it is women who have been perceived as holding the secret of life." Keller, *Secrets of Life*, 40.

36. See Keller, *Secrets of Life*, 41; Mary Jacobus, "Is There a Woman in This Text?" *New Literary History*, vol. 14, no. 1 (1982): 117–41.

37. Keller, *Secrets of Life*, 52–4.

38. Andrew Benjamin argues that what is assumed, even if not acknowledged, in the conception of difference assigned by Agamben to "bare life" is a primordial relatedness with the ineliminable specific other. He writes: "The excluded bear the mark not just of exclusion—a mark that could be no mark at all—but also the link between their particularity and exclusion." In "Particularity and Exception: On Jews and Animals," *The South Atlantic Quarterly*, vol. 107, no. 1 (Winter 2008): 71–87.

39. Watson, *DNA: The Secret of Life*, 106–7.

40. Watson, *DNA: The Secret of Life*, 58.

41. Ruth Hubbard and Elijah Wald explain in *Exploding the Gene Myth* that often it does not matter if you are a carrier of a particular gene, as, with diseases such as cystic fibrosis or Tay–Sachs, when recessive genes are involved, you need two gene carriers to procreate for the disease to develop.

42. Gibbs, "The Secret of Life," 43.

43. See Brenda Maddox, *Rosalind Franklin: The Dark Lady of DNA* (New York: HarperCollins, 2002), and Watson, *DNA: The Secret of Life*, 46–57.

44. Gilles Deleuze and Felix Guattari, *A Thousand Plateaus: Capitalism and Schizophrenia*, trans. Brian Massumi (London: The Athlone Press, 1998), 197.

45. I have borrowed the term "precarious life" from Judith Butler, *Precarious Life: The Power of Mourning and Violence* (London and New York: Verso, 2004).

46. The *Muselmann*, literally a Muslim, the living dead, is a product of the concentration camp. It is a term prisoners attached to those who had lost all will to live, and whose bodies approximated a vegetative state. Agamben uses this term after Primo Levi. See Giorgio Agamben, *Remnants of Auschwitz: The Witness and the Archive*, trans. Daniel Heller-Roazen (New York: Zone Books, 1999), 44.

47. As Maria Hynes argues, "when we speak about life we invariably bring into play a certain metaphysics. Indeed, any claim to offer a purely materialist description of life merely fails to examine the metaphysical assumptions made about the way that matter relates to an incorporeal or ideal dimension." In "Rethinking Reductionism," *Culture Machine*, vol. 7 (2004): nonpag.

48. Judith Butler, *Undoing Gender* (New York and London: Routledge, 2004), 18.

49. See Giorgio Agamben, *The Open: Man and Animal*, trans. Kevin Attell (Stanford: Stanford University Press, 2004), 16.

50. Agamben, *The Open*, 38.

51. Franklin, "Life," 1350.

52. Clare Birchall, "Cultural Studies Confidential," *Cultural Studies*, vol. 21, no. 1 (January 2007): 5–21, 18.

53. Eugene Thacker, *Biomedia* (Minneapolis and London: University of Minnesota Press, 2004), 184.

54. Birchall, "Cultural Studies Confidential," 18.

55. Richard Doyle, *On Beyond Living: Rhetorical Transformations of the Life Sciences* (Stanford: Stanford University Press, 1997), 10.

56. If we agree that the law is not God-given or eternal, we must envisage a time when it did not exist. The law cannot therefore have legitimized the force (or violence) that established it and that brought into being its authority. For more on this, see Jacques Derrida, "Force of Law: 'The Mystical Foundation of Authority'," *Cardozo Law Review: Deconstruction and the Possibility of Justice*, vol. 11, nos. 5–6 (1990): 920–1045.

57. Butler, *Undoing Gender*, 4.

58. Stephen Holland, *Bioethics: A Philosophical Introduction* (Cambridge: Polity, 2003), 1.

59. Emmanuel Levinas, *Ethics and Infinity: Conversations with Philippe Nemo* (Pittsburgh: Duquesne University Press, 1985), 74.

60. Emmanuel Levinas, *Totality and Infinity: An Essay on Exteriority*, trans. Alphonso Lingis (Pittsburgh: Duquesne University Press, 1969), 57–8.

61. Levinas, *Ethics and Infinity*, 78.

62. Butler, *Undoing Gender*, 19.

63. Levinas, *Ethics and Infinity*, 79.

64. Watson, *DNA: The Secret of Life*, 431.

65. Watson, *DNA: The Secret of Life*, 431.

Chapter 6

The last section of this chapter, "The Extra Ear of the Other: On Being-in-Difference," was originally published as an essay in a catalogue accompanying the exhibition of Stelarc's work (June 1–30, 2007), Experimental Art Foundation, Adelaide.

1. SymbioticA stands for the Art and Science Collaborative Research Laboratory at the School of Anatomy and Human Biology at the University of Western

Australia. Oron Catts and Ionat Zurr of SymbioticA also cofounded the Tissue Culture and Art Project, which uses tissue technologies as a medium of artistic expression. The aim of their projects, which include *Semi-Living Worry Dolls* (2000), *The Pig Wings Project* (2000), and *Disembodied Cuisine* (2003), is to interrogate humans' relationship with different "gradients of life" by growing a new class of objects, or beings—that of the semi-living.

2. Any such claims about the absolute novelty of a given art practice or medium should of course be treated with a certain degree of suspicion. In the article "Corporeal Mélange: Aesthetics and Ethics of Biomaterials in Stelarc and Nina Sellars's *Blender*," *Leonardo*, vol. 39, no. 5 (2006): 410–16, Julie Clarke traces a connection between Stelarc and Sellars's 2005 installation—which consisted of a large Plexiglas vessel in which subcutaneous fat, connective tissue, blood, and nerves extracted from both artists' bodies were being constantly blended—and Joseph Beuys's *Auschwitz* installation (1958) utilizing animal fat. According to Clarke, "Fat infiltrates and absorbs other materials and is a metaphor of transformation. Indeed, according to Beuys, fat relates to 'inner processes and feelings' [24]. It is in this sense that *Blender* enters into a dialogue with Beuys's use of fat, for in *Blender*, liquids and solids coalesce into what could be perceived as the chaotic embryonic state of a new body—one transformed by technological intervention" (414).

3. For a discussion of the conflicting responses to the work of Orlan and Stelarc, see Joanna Zylinska, " 'The Future . . . Is Monstrous': Prosthetics as Ethics," in *The Cyborg Experiments: The Extensions of the Body in the Media Age*, ed. Joanna Zylinska (London and New York: Continuum, 2002).

4. See http://www.critical-art.net/

5. Natalie Jeremijenko and Eugene Thacker, *Creative Biotechnology: A User's Manual* (Newcastle-upon-Tyne: Locus +, 2004), 10.

6. Jeremijenko and Thacker, *Creative Biotechnology*, 18.

7. Eduardo Kac, *Telepresence and Bio Art* (Ann Arbor: University of Michigan Press, 2005), 265–66.

8. Kac, *Telepresence and Bio Art*, 273.

9. Krzysztof Ziarek, *The Force of Art* (Stanford: Stanford University Press, 2004), 96–96.

10. Ziarek, *The Force of Art*, 96.

11. Joanna Zylinska, *The Ethics of Cultural Studies* (London and New York: Continuum, 2005), 6.

12. Wendy Brown, *Politics out of History* (Princeton and Oxford: Princeton University Press, 2001), 23.

13. I discuss the relationship between ethics, morality, and moralism in greater detail in Joanna Zylinska, "Cultural Studies and Ethics," in *New Cultural Studies: Adventures in Theory*, eds. Gary Hall and Clare Birchall (Edinburgh: Edinburgh University Press, 2006).

14. Jeremijenko and Thacker, *Creative Biotechnology*, 11.

15. Jeremijenko and Thacker, *Creative Biotechnology*, 11.

16. Ziarek, *The Force of Art*, 7.

17. Sarah Kember goes even further in criticizing what she sees as "false evolutionism" of Kac's GFP K-9 and GFP Bunny projects when she argues: "GFP is used so widely in test case transgenic experiments across art and science that it has given rise to what, for me, is the beguiling possibility that one day 'we' may all become green (if not little and men). And yet in both cases there is an (almost) inescapable sense of a template; of the containment of life-as-it-could-be by life-as-we-know-it; of accelerated evolution going nowhere fast and producing absolutely nothing new—perhaps, after Bergson, because of our fixation on change as the interval between entities and not the thing which links them together." Sarah Kember, "Creative Evolution? The Quest for Life (on Mars)," *Culture Machine*, InterZone (March 2006): nonpag.

18. Eugene Thacker, *The Global Genome: Biotechnology, Politics, and Culture* (Cambridge and London: MIT Press, 2005), 307.

19. Ziarek, *The Force of Art*, 3.

20. Thacker, *The Global Genome*, 307.

21. Ziarek, *The Force of Art*, 11.

22. "The discourse on, and of, (bio)ethics" does not just stand here for a general way of speaking about, or even theory of, ethics. As we have learned from Foucault and Laclau, "discourse" also articulates a certain materiality and is thus coexistent with the social. The discourse of bioethics therefore simultaneously *enacts* a bioethics.

23. Critical Art Ensemble, *Flesh Machine: Cyborgs, Designer Babies, and New Eugenic Consciousness* (Brooklyn, NY: Autonomedia, 1998), 60.

24. Critical Art Ensemble, *Flesh Machine*, 134. For an alternative view, see Catherine Waldby and Robert Mitchell, *Tissue Economies: Blood, Organs and Cell Lines in Late Capitalism* (Durham: Duke University Press, 2006).

25. http://www.critical-art.net/

26. Extensive documentation of all these projects, together with additional publications in the forms of books and frequently asked questions, is available on the CAE Web site: http://www.critical-art.net/

27. Critical Art Ensemble, *Flesh Machine*, 60.

28. For a more detailed explanation of the case, see the site of the CAE Defense Fund: http://www.caedefensefund.org/

29. Anna Munster foregrounds the importance of this work at the current conjuncture articulated as "war on terror," which has been used to define "global politics" post-9/11, by arguing: "It is essential to continue to support the cultural work of CAE's projects such as 'Free Range Grains' and the work-in-progress on U.S. germ warfare being carried out by Kurtz and others at the time of his

arraignment. Part of this support perhaps lies in separating, at least within the American mediascape—so bereft of serious and careful debate over terrorism and the war in Iraq—the work of art from the operations of terror. But at the same time we need to look for the broader cultural and political conditions today under which bioart might come to be considered as potential bioterrorism. We need to understand how a biopolitical logic modulates the politics of cultural and artistic production." Anna Munster, "Why Is Bioart *Not* Terrorism? Some Critical Nodes in the Networks of Informatic Life," *Culture Machine*, vol. 7 (2005): nonpag.

30. N. Katherine Hayles, *How We Became Posthuman* (Chicago and London: University of Chicago Press, 1999), 21.

31. Natalie Jeremijenko's *One Tree Project* (2003) involved planting 1,000 genetically identical clones of the same tree in various public places and monitoring their long-term growth.

32. Chapter 1 discusses different positions on bioethics in more detail.

33. The quote comes from a leaflet for the Macro/Micro Music Massage (MMMM) performance, which took place at Biofeel, the Biennial of Electronic Arts, in Perth in 2002. The public were invited to "join the process of sonic performance for cells in culture" by sitting in two "ButtVibe" vibrating chairs, which massaged the person sitting in one of the chairs in tune with the sound output sent from the vocalization of the person occupying the other chair.

34. Randy Kennedy, "The Artists in the Hazmat Suits," *The New York Times*, Sunday, July 3 (2005): 21.

35. John E. Mitchell, "Crossing Species Raises Issues of Ethics, Morals," *North Adams Transcript*, September 22 (2007): A1, A7; A7.

36. See Friedrich Wilhelm Nietzsche, *Twilight of the Idols, and, The anti-Christ*, trans. R. J. Hollingdale (Harmondsworth: Penguin, 1990).

37. Kac, *Telepresence and Bio Art*, 237–41.

38. Reodica's own work focuses on combining traditional art practices with emergent biotechnology and on exploring the ambiguities of the term "culture" as found both in human societies and in cellular organization. Her Living Sculpture series included the "hymNext Designer Hymen Project," "an installation in which unisex hymens [were] sculpted with living materials and the artist's own body cells into a variety of designs for the theoretical application upon the human body" in order to raise questions about the construction of sexuality and femininity in our culture. More information about Reodica's work can be found on her Web site: http://www.vivolabs.org.

39. Adam Zaretsky, "Workhorse Zoo Art and Bioethics Quiz," in *Biomediale: Contemporary Society and Genomic Culture*, edited and curated by Dmitry Bulatov (Kaliningrad: The National Publishing House "Yantarny Skaz," 2004): 321–34, 321.

40. Zaretsky, "Workhorse Zoo Art and Bioethics Quiz," 326–33.

41. Gilles Deleuze, *Spinoza: Practical Philosophy*, trans. Robert Hurley (San Francisco: City Lights Books, 1988), 27.

42. Jane Bennett, "The Agency of Assemblages and the North American Blackout," *Public Culture*, vol. 17, no. 3 (2005): 445–65, 446.

43. Jane Bennett, "The Agency of Assemblages," 447.

44. In another article, Bennett nevertheless admits that "even if nonhumans 'have' agentic capacity, its actualization and its strength will be affected by the character of *human* engagements with it." Jane Bennett, "Edible Material," *New Left Review*, vol. 45, May/June (2007): 133–45.

45. Bennett, "The Agency of Assemblages," 452.

46. This is a citation from Jacques Derrida, *Adieu: To Emmanuel Levinas*, trans. Pascale-Anne Brault and Michael Naas (Stanford: Stanford University Press, 1999), 23–24.

47. This citation, in turn, comes from Jacques Derrida, "I Have a Taste for the Secret," in *A Taste for the Secret*, eds. Jacques Derrida and Maurizio Ferraris (Cambridge: Polity, 2001), 56–57.

48. Gary Hall, *Digitize This Book!: The Politics of New Media, or Why We Need Open Access Now* (Minneapolis: University of Minnesota Press, 2008), 210–11.

49. Bennett, "The Agency of Assemblages," 457.

50. This Derrida-indebted definition of politics has been put forward by Ernesto Laclau and Chantal Mouffe. See Chantal Mouffe, *The Democratic Paradox* (London: Verso, 2000), 130.

51. Bennett, "The Agency of Assemblages," 464.

52. Deleuze, *Spinoza*, 22.

53. See Bennett, "The Agency of Assemblages," 446.

54. Bennett, "The Agency of Assemblages," 448.

55. Bennett, "The Agency of Assemblages," 452.

56. Zylinska, "The Future . . . Is Monstrous."

57. Jacques Derrida, *The Ear of the Other: Otobiography, Transference, Translation*, with Christie McDonald (New York: Schocken Books, 1985), 32–33; see also Nicholas Royle, *The Uncanny* (Manchester: Manchester University Press, 2003), 64.

58. Having had an ear-shaped Medpor scaffold implanted on his left arm in August 2006 which raises the skin and provides definition for an ear-like shape, the artist is currently working toward further surgeries. These will involve the construction of an earlobe and the implantation of a microphone into his arm.

59. Conversation with Stelarc, February 2007.

60. Paul Virilio, *Art and Fear* (London and New York: Continuum, 2003), 43, 93.

61. Bernard Stiegler, *Technics and Time, 1: The Fault of Epimetheus*, trans. Richard Beardsworth and George Collins (Stanford: Stanford University Press, 1998), 21.

62. Stiegler, *Technics and Time*, 141.

63. Stiegler, *Technics and Time*, 141.

64. Some of the ideas concerning Stiegler included in this paragraph come from the chapter "*Bio*-ethics and Cyberfeminism" in my book, *The Ethics of Cultural Studies*.

65. Brain Massumi, "The Evolutionary Alchemy of Reason," in *Stelarc: The Monograph*, ed. Marquard Smith (Cambridge and London: MIT Press, 2005), 179.

66. Massumi, "The Evolutionary Alchemy of Reason," 152.

67. Jane Goodall, "The Will to Evolve," in *Stelarc: The Monograph*, 8.

68. Goodall, "The Will to Evolve," 4–6.

69. Kember, "Creative Evolution?," nonpag. Kember notices, however, that "Where change itself may be accidental (since mutation works in different directions in different members of a species), the *tendency to change* is not. There is, for Bergson, 'an original impetus' of life which passes from one generation of germs to the next."

70. Arthur and Marilouise Kroker, "We Are All Stelarcs Now," in *Stelarc: The Monograph*.

Conclusion

1. Jacques Derrida, *Of Hospitality: Anne Dufourmantelle Invites Jacques Derrida to Respond*, trans. Rachel Bowlby (Stanford: Stanford University Press, 2000), 77.

Bibliography

Adami, C. (2002) "What Is Complexity?," *BioEssays*, vol. 24, no. 12: 1085–94.

Agamben, Giorgio (1998) *Homo Sacer: Sovereign Power and Bare Life*, trans. Daniel Heller-Roazen. Stanford: Stanford University Press.

Agamben, Giorgio (1999) *Remnants of Auschwitz: The Witness and the Archive*, trans. Daniel Heller-Roazen. New York: Zone Books.

Agamben, Giorgio (2004) "No to Biopolitical Tattooing," http://www.truthout .org/cgi-bin/artman/exec/view.cgi/4/3249

Agamben, Giorgio (2004) *The Open: Man and Animal*, trans. Kevin Attell. Stanford: Stanford University Press.

Agamben, Giorgio (2005) *State of Exception*, trans. Kevin Attell. Chicago: University of Chicago Press.

Badiou, Alain (2001) *Ethics: An Essay on the Understanding of Evil*, trans. Peter Hallward. London and New York: Verso.

Badmington, Neil (2006) "Cultural Studies and the Posthumanities," in *New Cultural Studies: Adventures in Theory*, eds. Gary Hall and Clare Birchall. Edinburgh: Edinburgh University Press.

Barker, Joanne (2004) "The Human Genome Diversity Project: 'Peoples,' 'Populations' and the Cultural Politics of Identification," *Cultural Studies*, vol. 18, no. 4: 571–606.

Beauchamp, Tom, and Childress, James F. (1979) *Principles of Biomedical Ethics*. Oxford: Oxford University Press.

Bennett, Jane (2005) "The Agency of Assemblages and the North American Blackout," *Public Culture*, vol. 17, no. 3: 445–65.

Bennett, Jane (2007) "Edible Material," *New Left Review*, vol. 45, May/June: 133–45.

Birchall, Clare (2007) "Cultural Studies Confidential," *Cultural Studies*, vol. 21, no. 1 (January): 5–21.

Borch-Jacobsen, Mikkel (1991) *Lacan: The Absolute Master*, trans. Douglas Brick. Stanford: Stanford University Press.

Braidotti, Rosi (2006) *Transpositions: On Nomadic Ethics*. Cambridge: Polity Press.

Broeckmann, Andreas (2005) "Introduction," Catalogue for *BASICS: Transmediale 05: International Media Art Festival Berlin*.

Brown, Wendy (2001) *Politics out of History*. Princeton and Oxford: Princeton University Press.

Butler, Judith (1997) *Excitable Speech: A Politics of the Performative*. New York and London: Routledge.

Butler, Judith (2004) *Precarious Life: The Power of Mourning and Violence*. London and New York: Verso.

Butler, Judith (2004) *Undoing Gender*. New York and London: Routledge.

Campbell, David, and Shapiro, Michael J. (eds.) (1999) *Moral Spaces: Rethinking Ethics and World Politics*. Minneapolis: University of Minnesota Press.

Cascais, Fernando (2003) "Bioethics: A Tentative Balance," *Studia Bioethica*, vol. 1, no. 1 (December): 25–34.

Chun, Wendy Hui Kyong (2006) "Introduction: Did Somebody Say New Media?," in *New Media, Old Media*, eds. Wendy Hui Kyong Chun and Thomas Keenan. New York and London: Routledge.

Clarke, Julie (2006) "Corporeal Mélange: Aesthetics and Ethics of Biomaterials in Stelarc and Nina Sellars's *Blender*," *Leonardo*, vol. 39, no. 5: 410–16.

Cohen, Kris R. (2006) "A Welcome for Blogs," *Continuum: Journal for Media & Culture Studies*, vol. 20, no. 2 (June): 161–73.

Colebrook, Claire (2002) *Gilles Deleuze*. London and New York: Routledge.

Colebrook, Claire (2007) "Agamben, Potentiality, and Life," *The South Atlantic Quarterly*, vol. 107, no. 1 (Winter): 107–20.

Critical Art Ensemble (1998) *Flesh Machine: Cyborgs, Designer Babies, and New Eugenic Consciousness*. Brooklyn, NY: Autonomedia.

Davis, Todd F., and Womack, Kenneth (eds.) (2001) *Mapping the Ethical Turn: A Reader in Ethics, Culture, and Literary Theory*. Charlottesville: University of Virginia Press.

Deleuze, Gilles (1988) *Spinoza: Practical Philosophy*, trans. Robert Hurley. San Francisco: City Lights Books.

Deleuze, Gilles, and Guattari, Felix (1988) *A Thousand Plateaus: Capitalism and Schizophrenia*, trans. Brian Massumi. London: The Athlone Press.

Derrida, Jacques (1976) *Of Grammatology*, trans. Gayatri Chakravorty Spivak. Baltimore and London: The Johns Hopkins University Press.

Derrida, Jacques (1978) "Violence and Metaphysics," in *Writing and Difference*, trans. Alan Bass. London and Henley: Routledge & Kegan Paul.

Derrida, Jacques (1985) *The Ear of the Other: Otobiography, Transference, Translation*, with Christie McDonald. New York: Schocken Books.

Derrida, Jacques (1986) "But, Beyond . . . (Open Letter to Anne McClintock and Rob Nixon)," *Critical Inquiry*, vol. 13, no. 1 (Autumn): 155–70.

Derrida, Jacques (1988) *Limited Inc.* Evanston: Northwestern University Press.

Derrida, Jacques (1990) "Force of Law: 'The Mystical Foundation of Authority,'" *Cardozo Law Review: Deconstruction and the Possibility of Justice*, vol. 11, no. 5–6: 920–1045.

Derrida, Jacques (1995) "'There is No *One* Narcissism' (Autobiophotographies)," in Jacques Derrida, *Points . . . Interview, 1974–1994*, ed. Elisabeth Weber. Stanford: Stanford University Press.

Derrida, Jacques (1999) *Adieu: To Emmanuel Levinas*, trans. Pascale-Anne Brault and Michael Naas. Stanford: Stanford University Press.

Derrida, Jacques (1999) "Hospitality, Justice and Responsibility: A Dialogue with Jacques Derrida," in *Questioning Ethics: Contemporary Debates in Philosophy*, eds. Richard Kearney and Mark Dooley. London and New York: Routledge.

Derrida, Jacques (2000) *Of Hospitality: Anne Dufourmantelle Invites Jacques Derrida to Respond*, trans. Rachel Bowlby. Stanford: Stanford University Press.

Derrida, Jacques (2001) "I Have a Taste for the Secret," in *A Taste for the Secret*, eds. Jacques Derrida and Maurizio Ferraris. Cambridge: Polity.

Derrida, Jacques (2002) "The Animal That Therefore I Am (More to Follow)," trans. David Wills, *Critical Inquiry*, vol. 28 (Winter): 369–418.

Deutscher, Penelope (2007) "The Inversion of Exceptionality: Foucault, Agamben, and 'Reproductive Rights,'" *The South Atlantic Quarterly*, vol. 107, no. 1 (Winter): 55–70.

Devenney, Mark, and Zylinska, Joanna (eds.) (2001) "Cultural Studies: Between Politics and Ethics" (special issue), *Strategies: Journal of Theory, Culture and Politics*, vol. 14, no. 2.

Diprose, Rosalyn (1994) *The Bodies of Women: Ethics, Embodiment and Sexual Difference*. London and New York: London.

Diprose, Rosalyn (1995) "A 'Genethics' That Makes Sense," in *Biopolitics: A Feminist and Ecological Reader on Biotechnology*, eds. Vandana Shiva and Ingunn Moser. London and Atlantic Highlands, NJ: Third World Network and Zed Books.

Diprose, Rosalyn (2002) *Corporeal Generosity: On Giving with Nietzsche, Merleau-Ponty, and Levinas*. Albany: SUNY Press.

Doyle, Richard (1997) *On Beyond Living: Rhetorical Transformations of the Life Sciences*. Stanford: Stanford University Press.

Elliott, Carl (1999) *A Philosophical Disease: Bioethics, Culture and Identity*. New York and London: Routledge.

Elliott, Carl (2003) "Not-So-Public Relations: How the Drug Industry Is Branding Bioethics," *Slate*, December 15.

Elliott, Carl (2004) *Better than Well: American Medicine Meets the American Dream*. New York and London: W.W. Norton & Company.

Finnis, John (1999) "Abortion and Health Care Ethics," in *Bioethics: An Anthology*, eds. Helga Kuhse and Peter Singer. Oxford: Blackwell.

Foucault, Michel (1984) *The History of Sexuality: An Introduction*, trans. Robert Hurley. Harmondsworth: Penguin Books. (Originally published in 1976)

Foucault, Michel (1990) "Technologies of the Self," in *Technologies of the Self*, eds. Luther H. Martin et al. London: Tavistock Books.

Foucault, Michel (1997) *Ethics: Subjectivity and Truth*, ed. Paul Rabinow. London: Allen Lane.

Foucault, Michel (2003) *"Society Must Be Defended": Lectures at the Collège de France, 1975–76*, trans. David Macey. London: Allen Lane.

Foucault, Michel (2005) *The Hermeneutics of the Subject: Lectures at the Collège de France 1981–82*, eds. Frederic Gros and François Ewald. Palgrave Macmillan: Basingstoke.

Frank, Arthur W., and Jones, Therese (2003) "Bioethics and the Later Foucault," *Journal of Medical Humanities*, vol. 24, nos. 3/4 (Winter): 179–86.

Franklin, Sarah (1995) "Life," in *The Encyclopedia of Bioethics*, ed. W. T. Reich, 5 vols. New York: Macmillan Library Reference USA.

Franklin, Sarah, Lury, Celia, and Stacey, Jackie (2000) *Global Nature, Global Culture*. London, Thousand Oaks, New Delhi: Sage.

Garber, Marjorie, Hanssen, Beatrice, and Walkowitz, Rebecca L. (eds.) (2000) *The Turn to Ethics*. New York and London: Routledge.

Gibbs, Nancy (2003) "The Secret of Life," *Time*, February 17: 42.

Gilman, Sander, L. (2000) *Making the Body Beautiful: A Cultural History of Aesthetic Surgery*. Princeton: Princeton University Press.

Glendinning, Simon (2007) *In the Name of Phenomenology*. London and New York: Routledge.

Goetz, Thomas (2007) "Your DNA Decoded," *Wired* 15.2: 256–65, 283.

Goodall, Jane (2005) "The Will to Evolve," in *Stelarc: The Monograph*, ed. Marquard Smith. Cambridge and London: MIT Press.

Gregg, Melissa (2006) "Feeling Ordinary: Blogging as Conversational Scholarship," *Continuum: Journal for Media & Cultural Studies*, vol. 20, no. 2 (June): 147–60.

Hall, Christian (2006) "An Interview with Alan Levy," *net*, Issue 157 (December): 34–6.

Hall, Gary (2008) *Digitize This Book!: The Politics of New Media, or Why We Need Open Access Now*. Minneapolis: University of Minnesota Press.

Hansen, Mark B. N. (2003) *New Philosophy for New Media*. Cambridge and London: MIT Press.

Haraway, Donna J. (1989) *Primate Visions: Gender, Race, and Nature in the World of Modern Science*. New York and London: Routledge.

Haraway, Donna J. (1991) *Simians, Cyborgs and Women: the Reinvention of Nature*. London: Free Association Books.

Haraway, Donna J. (1997) *Modest_Witness@Second_Millennium. FemaleMan©_Meets_OncoMouse™*. New York and London: Routledge.

Haraway, Donna (2003) *The Companion Species Manifesto: Dogs, People, and Significant Otherness*. Chicago: Prickly Paradigm Press.

Hardt, Michael, and Negri, Antonio (2000) *Empire*. Cambridge: Harvard University Press.

Harris, John (2007) *Enhancing Evolution: The Ethical Case for Making Better People*. Princeton and Oxford: Princeton University Press.

Hayles, N. Katherine (1995) "Making the Cut: The Interplay of Narrative and System, or What Systems Theory Can't See," *Cultural Critique*, no. 30, *The Politics of Systems and Environments*, part I (Spring): 71–100.

Hayles, N. Katherine (1999) *How We Became Posthuman*. Chicago and London: University of Chicago Press.

Hayles, N. Katherine (2002) *Writing Machines*. Cambridge and London: MIT Press.

Heidegger, Martin (1977) "The Question Concerning Technology," in *The Question Concerning Technology and Other Essays*, trans. William Lovitt. New York: Harper & Row.

Holland, Stephen (2003) *Bioethics: A Philosophical Introduction*. Cambridge: Polity.

Hubbard, Ruth, and Wald, Elijah (1997) *Exploding the Gene Myth*. Boston: Beacon Press.

Hynes, Maria (2004) "Rethinking Reductionism," *Culture Machine*, vol. 7, nonpag.

Jacobus, Mary (1982) "Is There a Woman in this Text?," *New Literary History*, vol. 14, no. 1: 117–41.

Jeremijenko, Natalie, and Thacker, Eugene (2004) *Creative Biotechnology: A User's Manual*. Newcastle-upon-Tyne: Locus +.

Jonsen, Albert R. (1998) *The Birth of Bioethics*. New York and Oxford: Oxford University Press.

Kac, Eduardo (2005) *Telepresence and Bio Art*. Ann Arbor: University of Michigan Press.

Kay, Lily E. (2000) *Who Wrote the Book of Life? A History of the Genetic Code*. Stanford: Stanford University Press.

Kearney, Richard, and Rainwater, Mara (eds.) (2005) *The Continental Philosophy Reader*. New York and London: Routledge.

Keller, Evelyn Fox (1992) *Secrets of Life, Secrets of Death: Essays on Gender, Language and Science*. New York and London: Routledge.

Kember, Sarah (2003) *Cyberfeminism and Artificial Life*. London and New York: Routledge.

Kember, Sarah (2006) "Creative Evolution? The Quest for Life (on Mars)," *Culture Machine*, InterZone (March): nonpag.

Kennedy, Randy (2005) "The Artists in the Hazmat Suits," *The New York Times*, Sunday, July 3: 21.

Kroker, Arthur, and Kroker, Marilouise (2005) "We Are All Stelarcs Now," in *Stelarc: The Monograph*, ed. Marquard Smith. Cambridge and London: MIT Press.

Kuhse, Helga, and Singer, Peter (1999) "Introduction," in *Bioethics: An Anthology*, eds. Helga Kuhse and Peter Singer. Oxford: Blackwell.

Lanchester, John (2006) "A Bigger Bang," *The Guardian Weekend*, November 4: 17–36.

Lasch, Christopher (1979) *The Culture of Narcissism: American Life in an Age of Diminishing Expectations*. London: Norton.

Lazzarato, Maurizio (no date) "From Biopower to Biopolitics," trans. Ivan A. Ramirez. Available at: http://www.geocities.com/immateriallabour/lazzarato-from-biopower-to-biopolitics.html. Accessed on May 31, 2007.

Lemonick, Michael L. (2003) "A Twist of Fate," *Time*, February 17: 48.

Levinas, Emmanuel (1969) *Totality and Infinity: An Essay on Exteriority*, trans. Alphonso Lingis. Pittsburgh: Duquesne University Press.

Levinas, Emmanuel (1985) *Ethics and Infinity: Conversations with Philippe Nemo*. Pittsburgh: Duquesne University Press.

Levinas, Emmanuel (1989) "Ethics as First Philosophy," in *The Levinas Reader*, ed. Sean Hand. Oxford: Blackwell.

Levinas, Emmanuel (1990) "The Name of a Dog, or Natural Rights," in *Difficult Freedom: Essays on Judaism*, trans. Sean Hand. London: The Athlone Press.

Levinas, Emmanuel (1998) *Otherwise Than Being: Or Beyond Essence*, trans. Alphonso Lingis. Pittsburgh: Duquesne University Press.

Levinas, Emmanuel, and Kearney, Richard (1986) "Dialogue with Emmanuel Levinas," in *Face to Face with Levinas*, ed. Richard A. Cohen. Albany: SUNY Press.

Lister, Martin, et al. (2003) *New Media: A Critical Introduction*. London and New York: Routledge.

Llewelyn, John (1991) "Am I Obsessed by Bobby? (Humanism and the Other Animal)," in *Re-reading Levinas*, eds. Robert Bernasconi and Simon Critchley. Bloomington and Indianapolis: Indiana University Press.

Mackenzie, Adrian (1996) "'God Has No Allergies': Immanent Ethics and the Simulacra of the Immune System," *Postmodern Culture*, vol. 6, no. 2 (January).

Mackenzie, Adrian (2002) *Transductions: Bodies and Machines at Speed*. Continuum: London and New York.

Maddox, Brenda (2002) *Rosalind Franklin: The Dark Lady of DNA*. New York: HarperCollins.

Marchitello, Howard (ed.) (2001) *What Happens to History: The Renewal of Ethics in Contemporary Thought*. New York and London: Routledge.

Marquis, Don (1999) "Why Abortion Is Immoral," in *Bioethics: An Anthology*, eds. Helga Kuhse and Peter Singer. Oxford: Blackwell.

Massumi, Brain (2005) "The Evolutionary Alchemy of Reason," in *Stelarc: The Monograph*, ed. Marquard Smith. Cambridge and London: MIT Press.

Maturana, Humberto R., and Varela, Francisco J. (1992) *The Tree of Knowledge: The Biological Roots of Human Understanding*, trans. Robert Paolucci. Boston: Shambhala.

Miah, Andy (2004) *Genetically Modified Athletes: Biomedical Ethics, Gene Doping and Sport*. London and New York: Routledge.

Miah, Andy (2005) "Genetics, Cyberspace and Bioethics: Why Not a Public Engagement with Ethics?," *Public Understanding of Science*, vol. 14, no. 4: 409–21.

Mills, Catherine (2003) "An Ethics of Bare Life: Agamben on Witnessing," *Borderlands*, vol. 3, no. 1: nonpag.

Mills, Catherine (2004) "Agamben's Messianic Politics: Biopolitics, Abandonment and Happy Life," *Contretemps*, vol. 5 (December): 42–62.

Mitchell, John E. (2005) "Crossing Species Raises Issues of Ethics, Morals," *North Adams Transcript*, September 22: A1, A7.

Mouffe, Chantal (2000) *The Democratic Paradox*. London: Verso.

Muller, Adam, and Livingston, Paisley (1995) "Realism/Anti-Realism: A Debate," *Cultural Critique*, no. 30, *The Politics of Systems and Environments*, part I (Spring): 15–32.

Munster, Anna (2005) "Why Is Bioart *Not* Terrorism? Some Critical Nodes in the Networks of Informatic Life," *Culture Machine*, vol. 7: nonpag.

Nelkin, Dorothy, and Lindee, Susan M. (1995) *The DNA Mystique: The Gene as a Cultural Icon*. New York: W. H. Freeman and Company.

Nietzsche, Friedrich Wilhelm (1990) *Twilight of the Idols, and, The anti-Christ*, trans. R. J. Hollingdale. Harmondsworth: Penguin.

Norris, Andrew (2005) "The Exemplary Exception: Philosophical and Political Decision on Giorgio Agamben's *Homo Sacer*," in *Politics, Metaphysics and Death: Essays on Giorgio Agamben's* Homo Sacer, ed. Andrew Norris. Durham and London: Duke University Press.

O'Leary, Timothy (2002) *Foucault and the Art of Ethics*. London and New York: Continuum.

O'Neill, Onora (1993) "Kantian Ethics," in *A Companion to Ethics*, ed. Peter Singer. Oxford: Blackwell.

Patton, Paul, and Protevi, John (2003) "Introduction," in *Between Deleuze and Derrida*, eds. Paul Patton and John Protevi. London and New York: Continuum.

Poster, Mark (2006) *Information Please: Culture and Politics in the Age of Digital Machines*. Durham and London: Duke University Press.

Poster, Mark (2008) "Swan's Way," *Configurations*, vol. 15 no. 2 (December).

Potter, Van Rensselaer (1970) "Bioethics, the Science of Survival," *Perspectives in Biology and Medicine*, vol. 14: 127–53.

Potter, Van Rensselaer (1971) *Bioethics: Bridge to the Future*. Englewood Cliffs, NJ: Prentice-Hall, Inc.

Purdy, Laura M. (1999) "Are Pregnant Women Fetal Containers?," in *Bioethics: An Anthology*, eds. Helga Kuhse and Peter Singer. Oxford: Blackwell.

Quastler, Henry (1956) "The Status of Information Theory in Biology," in *Symposium on Information Theory in Biology*, ed. H. P. Yockey. New York: Pergamon Press.

Rabinow, Paul (1997) "Introduction" to Michel Foucault, *Ethics: Subjectivity and Truth*, ed. Paul Rabinow. London: Allen Lane.

Rasch, William, and Wolfe, Cary (1995) "Introduction: The Politics of Systems and Environments," *Cultural Critique*, no. 30, *The Politics of Systems and Environments*, part I (Spring): 5–13.

Reed, Adam (2005) " 'My Blog Is Me': Texts and Persons in UK Online Journal Culture (and Anthropology)," *Ethnos*, vol. 70, no. 2 (June): 220–42.

Reich, Warren Thomas (1994) "The Word 'Bioethics': Its Birth and the Legacies of Those Who Shaped It," *Kennedy Institute of Ethics Journal*, vol. 4, no. 4: 319–35.

Reich, Warren Thomas (1995) "The Word 'Bioethics': The Struggle over Its Earliest Meanings," *Kennedy Institute of Ethics Journal*, vol. 5, no. 1: 19–34.

Rifkin, Jeremy (1998) *The Biotech Century*. New York: Tarcher.

Rose, Nikolas (2001) "The Politics of Life Itself," *Theory, Culture and Society*, vol. 18, no. 6: 1–30.

Ross, Alison (2007) "Introduction," *The South Atlantic Quarterly*, vol. 107, no. 1 (Winter): 1–13.

Royle, Nicholas (2003) *The Uncanny*. Manchester: Manchester University Press.

Saper, Craig (2006) "Blogademia," *Reconstruction*, vol. 6, no. 4: nonpag.

Schroedinger, Erwin (1944) *What Is Life? The Physical Aspect of the Living Cell*. Cambridge: Cambridge University Press.

Shildrick, Margrit (1997) *Leaky Bodies and Boundaries: Feminism, Postmodernism and (Bio)ethics*. London and New York: Routledge.

Shildrick, Margrit (2005) "Beyond the Body of Bioethics: Challenging the Conventions," in *Ethics of the Body: Postconventional Challenges*, eds. Margrit Shildrick and Roxanne Mykytiuk. Cambridge and London: MIT Press.

Shiva, Vandana (1997) *Biopiracy: The Plunder of Nature and Knowledge*. Cambridge: South End Press.

Singer, Peter (1994) *Rethinking Life and Death: The Collapse of Our Traditional Ethics*. Oxford: Oxford University Press.

Singer, Peter (2004) *The President of Good and Evil: The Ethics of George W. Bush*. London: Granta.

Singer, Peter, and Mason, Jim (2006) *Eating: What We Eat and Why It Matters*. London: Arrow Books.

Smith, Daniel W. (2003) "Deleuze and Derrida, Immanence and Transcendence: Two Directions in Recent French Thought," in *Between Deleuze and Derrida*, eds. Paul Patton and John Protevi. London and New York: Continuum.

Squier, Susan Merrill (2004) *Liminal Lives: Imagining the Human at the Frontiers of Biomedicine*. Durham and London: Duke University Press.

Stangroom, Jeremy (2005) "Darwinism and Genes: In Conversation with Steve Jones," in *What Scientists Think*. London and New York: Routledge: 1–22.

Stelarc (2005) "Prosthetic Head: Intelligence, Awareness and Agency," *CTheory*, nonpag.

Stiegler, Bernard (1998) *Technics and Time, 1: The Fault of Epimetheus*, trans. Richard Beardsworth and George Collins. Stanford: Stanford University Press.

Stiegler, Bernard (2003) "Technics of Decision: An Interview with Peter Hallward," trans. Sean Gaston. *Angelaki*, vol. 8, no. 2: 151–68.

Terranova, Tiziana (2004) *Network Culture: Politics for the Information Age*. London: Pluto Press.

Thacker, Eugene (2004) *Biomedia*. Minneapolis and London: University of Minnesota Press.

Thacker, Eugene (2005) "Nomos, Nosos and *Bios*," *Culture Machine*, vol. 7: nonpag.

Thacker, Eugene (2005) *The Global Genome: Biotechnology, Politics, and Culture*. Cambridge and London: MIT Press.

Thomson, Judith Jarvis (1999) "A Defense of Abortion," in *Bioethics: An Anthology*, eds. Helga Kuhse and Peter Singer. Oxford: Blackwell.

Tooley, Michael (1999) "Abortion and Infanticide," in *Bioethics: An Anthology*, eds. Helga Kuhse and Peter Singer. Oxford: Blackwell.

Van Dijck, José (2004) "Composing the self: Of diaries and lifelogs," *Fibreculture Journal*, no. 3: nonpag.

Varela, Francisco J. (1992) *Ethical Know-How: Action, Wisdom, and Cognition.* Stanford: Stanford University Press.

Virilio, Paul (2003) *Art and Fear*. London and New York: Continuum.

Wadiwel, Dinesh (2004) "Animal by Any Other Name? Patterson and Agamben Discuss Animal (and Human) Life," *Borderlands: e-journal*, vol. 3, no. 1: nonpag.

Waldby, Catherine (2000) *The Visible Human Project: Informatic Bodies and Posthuman Medicine*. London and New York: Routledge.

Waldby, Catherine, and Mitchell, Robert (2006) *Tissue Economies: Blood, Organs, and Cell Lines in Late Capitalism*. Durham: Duke University Press.

Wall, Thomas Carl (2005) "Bare Sovereignty: *Homo Sacer* and the Insistence of Law," in *Politics, Metaphysics and Death: Essays on Giorgio Agamben's* Homo Sacer, ed. Andrew Norris. Durham and London: Duke University Press.

Watson, James (2004) *DNA: The Secret of Life*. Arrow Books: London. (Originally published in 2003)

Weber, Samuel (1996) *Mass Mediauras: Form, Technics, Media*. Stanford: Stanford University Press.

Wegenstein, Bernadette (2006) *Getting under the Skin: The Body and Media Theory*. Cambridge and London: MIT Press.

Whitehouse, Peter J. (2002) "Van Rensselaer Potter: An Intellectual Memoir," *Cambridge Quarterly of Healthcare Ethics*, vol. 11: 331–34.

Wilson, E. O. (1975) *Sociobiology: The New Synthesis*. Cambridge, MA: Belknap Press.

Wolfe, Cary (1995) "In Search of Post-Humanist Theory: The Second-Order Cybernetics of Maturana and Varela," *Cultural Critique*, no. 30, *The Politics of Systems and Environments*, part I (Spring): 33–70.

Wolfe, Cary (2003) *Animal Rites: American Culture, the Discourse of Species, and Posthumanist Theory*. Chicago: University of Chicago Press.

Woodward, Kathleen (1983) "Cybernetic Modeling in Recent American Writing: A Critique," *North Dakota Quarterly*, vol. 51, no. 1 (Winter): 57–73.

Zaretsky, Adam (2004) "Workhorse Zoo Art and Bioethics Quiz," in *Biomediale: Contemporary Society and Genomic Culture*, ed. and curated by Dmitry Bulatov. Kaliningrad: The National Publishing House "Yantarny Skaz": 321–34.

Ziarek, Ewa Plonowska (2001) *An Ethics of Dissensus: Postmodernity, Feminism, and the Politics of Radical Democracy*. Stanford: Stanford University Press.

Ziarek, Ewa Plonowska (2007) "Bare Life on Strike: Notes on the Biopolitics of Race and Gender," *The South Atlantic Quarterly*, vol. 107, no. 1 (Winter): 89–105.

Ziarek, Krzysztof (2004) *The Force of Art*. Stanford: Stanford University Press.

Zylinska, Joanna (2002) " 'The Future . . . Is Monstrous': Prosthetics as Ethics," in *The Cyborg Experiments: The Extensions of the Body in the Media Age*, ed. Joanna Zylinska. London and New York: Continuum.

Zylinska, Joanna (2005) *The Ethics of Cultural Studies*. London and New York: Continuum.

Zylinska, Joanna (2006) "Cultural Studies and Ethics," in *New Cultural Studies: Adventures in Theory*, eds. Gary Hall and Clare Birchall. Edinburgh: Edinburgh University Press.

Zylinska, Joanna (2006) "Dogs R Us?" Review of Donna Haraway's *The Companion Species Manifesto*, *parallax*, vol. 12, no. 1: 129–31.

Index